SAS® 9.4 Intelligence Platform: Overview, Second Edition

D1532637

SAS® Documentation

The correct bibliographic citation for this manual is as follows: SAS Institute Inc. 2016. *SAS® 9.4 Intelligence Platform: Overview, Second Edition.* Cary, NC: SAS Institute Inc.

SAS® 9.4 Intelligence Platform: Overview, Second Edition

Copyright © 2016, SAS Institute Inc., Cary, NC, USA

ISBN 978-1-62960-089-5 (Hard copy)
ISBN 978-1-62960-090-1 (PDF)

Contents

Chapter 1
Value of the SAS Intelligence Platform

About the SAS Business Analytics Framework and the SAS Intelligence Platform

The SAS Business Analytics Framework encompasses a comprehensive set of business solutions, technologies, and services from SAS. Through this framework, organizations can address their most critical business issues and then add new functionality over time to enable continuous performance improvement. All of the functionality is available from one vendor and through one framework, thus reducing the total cost of ownership.

The technologies that provide the foundation for the SAS Business Analytics Framework, as well as for industry and line-of-business solutions offered by SAS, are delivered through the SAS Intelligence Platform. The SAS Intelligence Platform is a comprehensive, end-to-end infrastructure for creating, managing, and distributing enterprise intelligence. It includes tools and interfaces that enable you to do the following:

- extract data from a variety of operational data sources on multiple platforms, and build a data warehouse and data marts that integrate the extracted data

- store large volumes of data efficiently and in a variety of formats

- give business users at all levels the ability to explore data from the warehouse in a web browser, perform simple query and reporting functions, and view up-to-date results of complex analyses

- use high-end analytic techniques to provide capabilities such as predictive and descriptive modeling, forecasting, optimization, simulation, and experimental design

- centrally control the accuracy and consistency of enterprise data

With the SAS Intelligence Platform, you can implement an end-to-end intelligence infrastructure using software that is delivered, tested, and integrated by SAS. Using the tools provided in the SAS Intelligence Platform, you can create applications that reflect your unique business requirements and domain knowledge.

Building on the technologies in the SAS Intelligence Platform, SAS offers solutions for industries such as financial services, life sciences, health care, retail, and manufacturing, as well as line-of-business solutions in areas such as customer management, enterprise risk management, and financial management. The solutions incorporate predictive analytics, industry and domain expertise, and specialized data structures.

A complete set of documentation for deploying and administering the SAS Intelligence Platform is available at the following locations: SAS Intelligence Platform: Installation, Configuration, and Migration and SAS Intelligence Platform: Administration. The Help Center provides the primary administrative documentation for several software offerings, including the following:

- SAS BI Server

- SAS Enterprise BI Server

- SAS Data Management (Standard or Advanced) and other related offerings for data management, data quality, and data governance

The SAS Data Integration Server offering has been revised, and the SAS Enterprise Data Integration Server offering is available only through renewals.

Administrators who manage SAS products that use the metadata server should use the documentation at this site. Supplemental guides provide additional details for other products and domains.

Components of the SAS Intelligence Platform

Components Overview

The SAS Intelligence Platform includes components in the following categories:

Data Management
> The data management components enable you to consolidate and manage enterprise data from a variety of source systems, applications, and technologies. Components are provided to help you cleanse, migrate, synchronize, replicate, and promote your data. In addition, SAS offers data storage options that are optimized for analytical processing, enabling you to quickly analyze and report on large volumes of data. Metadata for all of your intelligence resources is stored centrally and controlled through a single management interface.

Business Intelligence
> The business intelligence components enable users with various needs and skill levels to create, produce, and share their own reports and analyses. Through easy-to-use interfaces, users can obtain their own answers to business questions. Meanwhile,

the information technology staff retains control over the quality and consistency of the data.

Advanced Analytics

SAS offers the richest and widest portfolio of analytic products in the software industry. The portfolio includes products for predictive and descriptive modeling, data mining, text analytics, forecasting, optimization, simulation, data visualization, model management, and experimental design. You can use any combination of these tools with the SAS Intelligence Platform to add extraordinary precision and insight to your reports and analyses.

The following sections describe in more detail the data management, business intelligence, and analytics components, as well as key supporting components.

Data Management Overview

Data Management

The software tools in the data management category enable you to consolidate and manage enterprise data from a variety of source systems, applications, and technologies. SAS provides access engines and interfaces to a wide variety of data sources, including the following:

- delimited files, SAS data sets, and relational database management system (RDMS) tables

- Hadoop file systems for handling big data

- application data from enterprise resource planning (ERP) and customer relationship management (CRM) systems

- message queuing platforms

- web services

- unstructured and semi-structured data

Data storage options include simple relational databases, a threaded multidimensional database that supports online analytical processing (OLAP), and relational storage with a threaded multiple input/output (I/O) subsystem for intensive use by focused applications.

Each of the data management solutions is described briefly in the following sections.

SAS Data Integration Studio

SAS Data Integration Studio is a visual design tool that enables you to consolidate and manage enterprise data from a variety of source systems, applications, and technologies. The software enables you to create jobs and process flows that extract, transform, and load data for use in data warehouses and data marts. You can also create processes that cleanse, migrate, synchronize, replicate, and promote data for applications and business services.

SAS Data Integration Studio is part of the following offerings: SAS Enterprise Data Integration Server (renewals only), SAS Data Integration Server, SAS Data Management Standard, and SAS Data Management Advanced.

For more information, see Chapter 6, "Clients in the SAS Intelligence Platform," on page 41.

SAS Data Loader for Hadoop

SAS Data Loader for Hadoop is a software offering that makes it easier to move, cleanse, and analyze data in Hadoop. It enables business users and data scientists to do self-service data preparation on a Hadoop cluster. Hadoop is highly efficient at storing and processing large amounts of data. However, moving, cleansing, and analyzing data in Hadoop can be labor intensive, and these tasks usually require specialized coding skills. As a result, business users and data scientists usually depend on IT personnel to prepare large Hadoop data sets for analysis. This technical overhead makes it harder to turn Hadoop data into useful knowledge.

SAS Data Loader for Hadoop provides a set of wizards, called directives, that help business users and data scientists perform the following tasks:

- copy data to and from Hadoop using parallel, bulk data transfer

- perform data integration, data quality, and data preparation tasks within Hadoop without writing complex MapReduce code

- minimize data movement for increased scalability, governance, and performance

- load data in memory to prepare it for high-performance reporting, visualization, or analytics

DataFlux Data Management Platform

Software in the DataFlux Data Management Platform enables you to discover, design, deploy, and maintain data across your enterprise in a centralized way. Data quality, data integration, and master data management are all provided under a unified user interface called DataFlux Data Management Studio. DataFlux Web Studio provides a web interface for managing a list of business data terms, for managing reference data, or for viewing exceptions to monitored business rules.

These DataFlux products work with SAS Data Integration Studio and are part of the SAS Data Management Standard and SAS Data Management Advanced offerings.

SAS Data Quality Server

SAS Data Quality Server works with SAS Data Integration Studio and the DataFlux Data Management Platform to analyze, cleanse, transform, and standardize your data. The language elements that make up the SAS Data Quality Server software form the basis of the data quality transformations in SAS Data Integration Studio.

SAS Data Quality Accelerator for Teradata

SAS Data Quality Accelerator for Teradata allows data quality functions to be invoked within the database, eliminating the need to move data off and back onto the database server as part of data quality processing.

SAS Data Surveyor for SAP

SAS Data Surveyor for SAP enables you to build SAS Data Integration Studio jobs that read data directly from SAP R/3 and SAP Business Warehouse systems.

Data Storage Options

The data storage options that can be used with the SAS Intelligence Platform include SAS data tables, parallel storage, multidimensional databases, Hadoop file systems, and third-party databases. These storage options can be used alone or in any combination. Metadata for your intelligence resources is stored centrally in the SAS Metadata Repository for use by all components of the intelligence platform.

Relational Storage: SAS Data Sets

You can use SAS data sets, the default SAS storage format, to store data of any granularity. The data values in a SAS data set are organized as a table of observations (rows) and variables (columns). A SAS data set also contains descriptor information such as the data types and lengths of the columns, as well as which SAS engine was used to create the data.

Access to Third-Party Databases: SAS/ACCESS

SAS/ACCESS provides interfaces to a wide range of relational, hierarchical, and network model databases. Examples include DB2, Oracle, SQL Server, Teradata, Hadoop, IBM Information Management System (IMS), and Computer Associates Integrated Database Management System (CA-IDMS). With SAS/ACCESS, SAS Data Integration Studio and other SAS applications can read, write, and update data regardless of which database and platform the data is stored on. SAS/ACCESS interfaces provide fast, efficient data loading and enable SAS applications to work directly from your data sources without making a copy.

High-Performance Computing: SAS In-Database

To support high-performance computing for complex, high-volume analytics, SAS In-Database enables certain data management, analytic, and reporting tasks to be performed inside the database. In-database technology minimizes the movement of data across the network, while enabling more sophisticated queries and producing results more quickly. This technology is available for several types of databases.

Multidimensional Storage: SAS OLAP Server

The SAS OLAP Server provides dedicated storage for data that has been summarized along multiple business dimensions. The server uses a threaded, scalable, and open technology and is especially designed for fast-turnaround processing and reporting.

A simplified ETL process enables you to build consistent OLAP cubes from disparate systems. A threaded query engine and parallel storage enable data to be spread across multiple-disk systems. Support is provided for multidimensional (MOLAP) and hybrid (HOLAP) data stores, as well as for open industry standards.

Parallel Storage: SAS Scalable Performance Data Engine and SAS Scalable Performance Data Server

The SAS SPD Engine and SAS SPD Server provide a high-speed data storage alternative for processing very large SAS data sets. They read and write tables that contain millions of observations, including tables that exceed the 2-GB size limit imposed by some operating systems. In addition, they provide the rapid data access that is needed to support intensive processing by SAS analytic software and procedures.

These facilities work by organizing data into a streamlined file format and then using threads to read blocks of data very rapidly and in parallel. The software tasks are performed in conjunction with an operating system that enables threads to execute on any of the CPUs that are available on a machine.

The SAS SPD Engine, which is included with Base SAS software, is a single-user data storage solution. The SAS SPD Server, which is available as a separate product, is a multi-user solution that includes a comprehensive security infrastructure, backup and restore utilities, and sophisticated administrative and tuning options.

Business Intelligence

The software tools in the business intelligence category address two main functional areas: information design, and self-service reporting and analysis.

The information design tools enable business analysts and information architects to organize data in ways that are meaningful to business users, while shielding the end users from the complexities of underlying data structures. These tools include the following products:

- SAS Information Map Studio enables analysts and information architects to create and manage information maps that contain business metadata about your data.

- SAS OLAP Cube Studio enables information architects to create cube definitions that organize summary data along multiple business dimensions.

The self-service reporting and analysis tools enable business users to query, view, and explore centrally stored information. Users can create their own reports, graphs, and analyses in the desired format and level of detail. In addition, they can find, view, and share previously created reports and analyses. The tools feature intuitive interfaces that enable business users to perform these tasks with minimal training and without the involvement of information technology staff.

The self-service reporting and analysis tools include the following products:

- SAS Web Report Studio is a web-based query and reporting tool that enables users at any skill level to create, view, and organize reports.

- SAS Information Delivery Portal provides a web-based, personalized workplace to help decision makers easily find the information that they need.

- SAS BI Portlets includes portlets, such as the SAS Stored Process Portlet and the SAS Report Portlet, that add value to the SAS Information Delivery Portal.

- SAS BI Dashboard enables SAS Information Delivery Portal users to create, maintain, and view dashboards to monitor key performance indicators that convey how well an organization is performing.

- SAS Add-In for Microsoft Office enables users to access SAS functionality from within Microsoft Office products.

- SAS Enterprise Guide is a project-oriented Windows application that enables users to create processes that include complex computations, business logic, and algorithms.

- SAS Mobile BI enables users to use mobile devices (iPad and Android) to view certain relational reports that have been created with SAS Web Report Studio.

As users create information maps, cubes, report definitions, portal content definitions, and stored processes, information about them is stored in the SAS Metadata Repository. Client applications and users can access these information assets on a need-to-know basis. Access is controlled through multilayered security that is enforced through the metadata.

For a description of each of the business intelligence tools, see Chapter 6, "Clients in the SAS Intelligence Platform," on page 41.

Analytics

SAS offers the richest and widest portfolio of analytic products in the software industry. The portfolio includes products for predictive and descriptive modeling, data mining, text analytics, forecasting, optimization, simulation, and experimental design. You can use any combination of these tools with the SAS Intelligence Platform to add precision and insight to your reports and analyses.

SAS software provides the following types of analytical capabilities:

- predictive analytics and data mining, to build descriptive and predictive models and deploy the results throughout the enterprise

- text analytics, to maximize the value of unstructured data assets

- dynamic data visualization, to enhance the effectiveness of analytics

- forecasting, to analyze and predict outcomes based on historical patterns

- model management, to streamline the process of creating, managing, and deploying analytic models

- operations research, to apply techniques such as optimization, scheduling, and simulation to achieve the best result

- quality improvement, to identify, monitor, and measure quality processes over time

- statistical data analysis, to drive fact-based decisions

Here are examples of analytic products:

- SAS Enterprise Miner enables analysts to create and manage data mining process flows. These flows include steps to examine, transform, and process data to create models that predict complex behaviors of economic interest. The SAS Intelligence Platform enables SAS Enterprise Miner users to centrally store and share the metadata for models and projects. In addition, SAS Data Integration Studio provides the ability to schedule data mining jobs.

- SAS Forecast Server enables organizations to plan more effectively for the future by generating large quantities of high-quality forecasts quickly and automatically. This solution includes the SAS High-Performance Forecasting engine, which selects the time series models, business drivers, and events that best explain your historical data, optimizes all model parameters, and generates high-quality forecasts. SAS Forecast Studio provides a graphical interface to these high-performance forecasting procedures.

- SAS Model Manager supports the deployment of analytical models into your operational environments. It enables registration, modification, tracking, scoring, and reporting on analytical models that have been developed for BI and operational applications.

- JMP is interactive, exploratory data analysis and modeling software for the desktop. JMP makes data analysis—and the resulting discoveries—visual and helps communicate those discoveries to others. JMP presents results both graphically and numerically. By linking graphs to each other and to the data, JMP makes it easier to see the trends, outliers, and other patterns that are hidden in your data.

Supporting Components

SAS Metadata Repository

Your information assets are managed in a common metadata layer called the SAS Metadata Repository.

This repository stores logical data representations of items such as libraries, tables, information maps, and cubes, thus ensuring central control over the quality and consistency of data definitions and business rules. The repository also stores information about system resources such as servers, the users who access data and metadata, and the rules that govern who can access what.

All of the data management and business intelligence tools read and use metadata from the repository and create new metadata as needed.

Administrative Interfaces

With SAS 9.4, SAS introduces SAS Environment Manager as the next-generation web-based interface for SAS administration. Based on the VMware Hyperic product, SAS Environment Manager incorporates the monitoring and managing of IT and SAS resources into an overall service-management strategy. It also enables you to perform SAS administration functions including authorization, user management, server management, library management, and backup. For more information, see "SAS Environment Manager" on page 45.

SAS Management Console provides a single interface through which system administrators can manage and monitor SAS servers, explore and manage metadata repositories, manage user and group accounts, and administer security.

Scheduling in SAS

Platform Suite for SAS is an optional product that provides enterprise-level scheduling capabilities in a single-server environment. Platform Suite for SAS is also included as part of the SAS Grid Manager product to enable distributed enterprise scheduling, workload balancing, and parallelized workload balancing. The components of Platform Suite for SAS include Process Manager, Load Sharing Facility (LSF), and Grid Management Services.

As an alternative, operating system services can be used to provide a basic level of scheduling for SAS jobs, and SAS in-process scheduling enables you to schedule jobs from certain web-based SAS applications.

Strategic Benefits of the SAS Intelligence Platform

Multiple Capabilities Integrated into One Platform

The SAS Intelligence Platform combines advanced SAS analytics, high-speed processing of large amounts of data, and easy-to-use query and reporting tools. The result is accurate, reliable, and fast information with which to make decisions.

You can build data warehouses, perform data mining, enable users to query data and produce reports from a web browser, and give users easy access to SAS processes that perform robust analytics.

The SAS Intelligence Platform provides all of this functionality in one centrally managed suite of products that are designed to work together seamlessly. This integration reduces the administration, management, and deployment costs that would be associated with providing multiple technologies to meet the needs of different users.

Consistency of Data and Business Rules

The SAS Intelligence Platform uses your organization's existing data assets, enabling you to integrate data from multiple database platforms and ERPs. Tools are provided to help ensure the reliability, consistency, and standardization of this data.

Users can choose from multiple tools with which to perform queries and produce reports. Since all of the tools access data through the same metadata representations, users throughout your enterprise receive consistent data. As a result, they can make decisions based on a common version of the truth.

Similarly, business logic, complex computations, and analytic algorithms can be developed once and stored centrally in SAS processes for all users to access. These processes, as well as the information in the SAS Metadata Repository, are controlled through multi-level security.

Fast and Easy Reporting and Analysis

The SAS Intelligence Platform's self-service reporting and analysis tools enable users across the enterprise to access and query data from virtually any data source. Any number of users can use wizards to create reports in the needed time frames, without waiting for support from information technology professionals. Through web-based interfaces, users can explore large volumes of multidimensional data quickly and interactively, from multiple perspectives and at multiple levels of detail.

The reporting and analysis tools hide complex data structures, so that average business users can perform queries without having to learn new skills. The intelligence storage options are optimized for analytical processing, enabling the reporting tools to quickly retrieve large volumes of data.

As a result of these reporting and analysis capabilities, everyone spends less time looking for answers and more time driving strategic decisions.

Analytics Available to All Users

SAS is the market leader in analytics. With the SAS Intelligence Platform, you can make the full breadth of SAS analytics available to users throughout the enterprise.

SAS analytics include algorithms for functions such as predictive and descriptive modeling, forecasting, optimization, simulation, and experimental design. You can now incorporate these capabilities into self-service reports and analyses, so that decision makers throughout your enterprise can benefit from the accuracy and precision of high-end analytics.

About SAS Visual Analytics

SAS Visual Analytics complements the capabilities of the SAS Intelligence Platform. SAS Visual Analytics is a high-performance, in-memory solution for exploring massive amounts of data very quickly. The offering includes a highly visual drag-and-drop interface that enables users to quickly identify patterns and trends in huge volumes of data and identify opportunities for further analysis. An easy-to-use visual design interface enables users to create reports or dashboards that can be viewed on a mobile device or on the web.

Using these tools, business users can explore data on their own to develop new insights, and analysts can speed up the modeling cycle by quickly identifying statistically relevant variables. A key component of SAS Visual Analytics is the SAS LASR Analytic Server, which quickly reads data into memory for fast processing and data visualization.

The SAS Visual Analytics infrastructure includes some of the same software components that are included in the SAS Intelligence Platform. However, SAS Visual Analytics is installed in a dedicated environment that includes specialized hardware and its own instances of SAS software and servers.

For more information, see http://www.sas.com/software/visual-analytics/overview.html.

About SAS Viya

SAS Viya is an open, cloud-enabled, in-memory engine on the SAS Platform. It provides elastic, scalable, and fault-tolerant processes to address your complex analytical challenges.

SAS9.4M5 supports sessions between SAS and the SAS Viya 3.2 Cloud Analytics Services (CAS) server. The CAS server can run on a single machine or as a distributed server on multiple machines. For both modes, the server is multi-threaded for high-performance analytics.

Chapter 2

Architecture of the SAS Intelligence Platform

Architecture Overview

The SAS Intelligence Platform architecture is designed to efficiently access large amounts of data, while simultaneously providing timely intelligence to a large number of users. The platform uses an n-tier architecture that enables you to distribute functionality across computer resources, so that each type of work is performed by the resources that are most suitable for the job.

You can easily scale the architecture to meet the demands of your workload. For a large company, the tiers can be installed across a multitude of machines with different operating systems. For prototyping, demonstrations, or very small enterprises, all of the tiers can be installed on a single machine.

The architecture consists of the following four tiers:

Data sources
    Data sources store your enterprise data. All of your existing data assets can be used, whether your data is stored in third-party database management systems, SAS tables, or enterprise resource planning (ERP) system tables.

SAS servers
    SAS servers perform SAS processing on your enterprise data. Several types of SAS servers are available to handle different workload types and processing intensities.

The software distributes processing loads among server resources so that multiple client requests for information can be met without delay.

Middle tier

The middle tier enables users to access intelligence data and functionality via a web browser. This tier provides web-based interfaces for report creation and information distribution, while passing analysis and processing requests to the SAS servers.

Clients

The client tier provides users with desktop access to intelligence data and functionality through easy-to-use interfaces. For most information consumers, reporting and analysis tasks can be performed with just a web browser. For more advanced design and analysis tasks, SAS client software is installed on users' desktops. Some support for mobile devices is also provided.

Note: The four tiers represent categories of software that perform similar types of computing tasks and require similar types of resources. The tiers do not necessarily represent separate computers or groups of computers.

The following diagram shows how the tiers interact, and the sections that follow describe each tier in more detail.

Figure 2.1 *Architecture of the SAS Intelligence Platform*

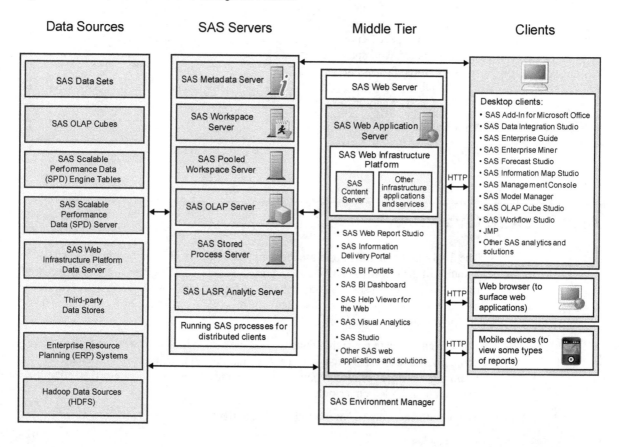

Data Sources

The SAS Intelligence Platform includes the following options for data storage:

- SAS data sets, which are analogous to relational database tables.

- SAS SPD Engine tables, which can be read or written by multiple threads.

- SAS SPD Server, which is available as a separate product.

- SAS OLAP cubes.

- The SAS Web Infrastructure Platform Data Server, which is the default location for middle-tier data such as alerts, comments, and workflows, as well as data for the SAS Content Server. The server is provided as an alternative to using a third-party DBMS. (The server cannot be used as a general-purpose data store.)

In addition, SAS provides products that enable you to access data in your existing third-party data stores and ERP systems, including:

- the SAS/ACCESS interfaces, which provide direct access to a variety of data stores. For a complete list, go to http://support.sas.com/software/products/access/.

- SAS Data Surveyor for SAP, which enables you to build SAS Data Integration Studio jobs that read data directly from SAP R/3 and SAP Business Warehouse systems.

For more information about data sources, see Chapter 3, "Data in the SAS Intelligence Platform," on page 17.

SAS Server Tier

SAS Server Tier Overview

The SAS server tier includes the SAS Metadata Server and several compute servers that execute SAS analytical and reporting processes for distributed clients. These servers are typically accessed either by desktop clients or by web applications that are running in the middle tier.

Note: In the SAS Intelligence Platform, the term server refers to a process or processes that wait for and fulfill requests from client programs for data or services. The term server does not necessarily refer to a specific computer, since a single computer can host one or more servers of various types.

SAS Metadata Server

The SAS Metadata Server controls access to a central repository of metadata that is shared by all of the SAS applications in the deployment. The SAS Metadata Server enables centralized control so that all users access consistent and accurate data. The metadata repository stores information about the following:

- the enterprise data sources and data structures that are accessed by SAS intelligence applications.

- the content that is created and used by SAS applications. This content includes information maps, OLAP cubes, report definitions, stored process definitions, and portal content definitions.

- the SAS and third-party servers that participate in the system.

- users and groups of users who are allowed to use the system. Users can be authenticated by the metadata server or by external systems such as the host environment, the web realm, and third-party databases.

- the levels of access that users and groups have to resources. This metadata-based authorization layer supplements protections from the host environment and other systems.

The SAS Intelligence Platform provides a central management tool—SAS Management Console—that you use to manage the metadata server and the metadata repository.

SAS OLAP Server

The SAS OLAP Server is a multidimensional data server that delivers pre-summarized cubes of data to business intelligence applications. The data is queried using the MDX (multidimensional expression) language.

This server is designed to reduce the load on traditional back-end storage systems by quickly delivering summarized views, regardless of the amount of data that underlies the summaries.

SAS Workspace Server

The SAS Workspace Server enables client applications to submit SAS code to a SAS session using an application programming interface (API). For example, when you use SAS Data Integration Studio to submit an ETL job for processing, the application generates the SAS code necessary to perform the processing and submits it to a workspace server. You can run as many instances of workspace servers as are needed to support your workload.

SAS Pooled Workspace Server

The SAS Pooled Workspace Server is a workspace server that uses server-side pooling. This configuration maintains a collection of reusable workspace server processes for clients, thus avoiding the overhead associated with creating a new process for each connection. Clients such as SAS Information Map Studio, SAS Web Report Studio, and the SAS Information Delivery Portal can use pooled workspace servers to query relational data.

SAS Stored Process Server

The SAS Stored Process Server executes and delivers results from SAS Stored Processes in a multi-client environment. A SAS Stored Process is a SAS program that is stored centrally and that can be executed by users and client programs on demand.

You can run as many instances of stored process servers as are needed to support your workload.

SAS Object Spawner

The SAS object spawner is a process that runs on workspace server, pooled workspace server, and stored process server host machines. It listens for requests for these servers, authenticates the requesting clients, and launches server processes as needed. In a pooled workspace server configuration, the object spawner maintains a collection of re-usable workspace server processes that are available for clients. If server load balancing is configured, the object spawner balances workloads between server processes. The object spawner connects to the metadata server to obtain information about the servers that it manages.

SAS LASR Analytic Server

The SAS LASR Analytic Server is an analytic platform that is provided with SAS Visual Analytics, SAS Visual Statistics, and other high-performance products. This secure, multi-user server provides concurrent access to data that is loaded into memory. The server can take advantage of a distributed computing environment by distributing data and the workload among multiple machines and performing massively parallel processing. The server processes client requests at extraordinarily high speeds due to the combination of hardware and software that is designed for rapid access to tables in memory.

Middle Tier

The middle tier of the SAS Intelligence Platform provides an environment in which the business intelligence web applications, such as SAS Web Report Studio and the SAS Information Delivery Portal, can execute. These products run in a web application server and communicate with the user by sending data to and receiving data from the user's web browser. The middle tier applications rely on servers on the SAS server tier to perform SAS processing, including data query and analysis.

The middle tier includes the following SAS software elements:

- the SAS Web Application Server and SAS Web Server

- SAS web applications, which can include SAS Web Report Studio, the SAS Information Delivery Portal, SAS BI Portlets, the SAS BI Dashboard, SAS Environment Manager, and other SAS products and solutions

- the SAS Web Infrastructure Platform, which includes the SAS Content Server and other infrastructure applications and services

For more information about the middle tier, see Chapter 5, "Middle-Tier Components of the SAS Intelligence Platform," on page 35.

Clients

The clients in the SAS Intelligence Platform provide web-based and desktop user interfaces to content and applications. SAS clients provide access to content, appropriate query and reporting interfaces, and business intelligence functionality for all of the information consumers in your enterprise, from the CEO to business analysts to customer service agents.

The following clients run on Windows desktops. Some of these clients are native Windows applications and others are Java applications.

- SAS Add-In for Microsoft Office

- SAS Data Integration Studio

- SAS Enterprise Guide

- SAS Enterprise Miner

- SAS Forecast Studio

- SAS Information Map Studio

- JMP (also available on Macintosh and Linux)

- SAS Management Console

- SAS Model Manager

- SAS OLAP Cube Studio

- SAS Workflow Studio

SAS Management Console is supported on all platforms except z/OS.

The following products require only a web browser to be installed on each client machine, with the addition of Adobe Flash Player for SAS BI Dashboard:

- SAS Information Delivery Portal

- SAS BI Dashboard

- SAS Environment Manager

- SAS Web Report Studio

- SAS Help Viewer for the Web

In addition, SAS Mobile BI enables users to use mobile devices (iPad and Android) to view certain relational reports that have been created with SAS Web Report Studio.

For more information about the clients, see Chapter 6, "Clients in the SAS Intelligence Platform," on page 41.

Chapter 3
Data in the SAS Intelligence Platform

Overview of Data Storage Options

In a SAS Intelligence Platform deployment, you can use one or more of these data storage options:

- default SAS storage in the form of SAS tables

- third-party relational, hierarchical, and network model database tables

- parallel storage from the SAS Scalable Performance Data Engine (SPD Engine) and the SAS Scalable Performance Data Server (SPD Server)

- multidimensional databases (cubes)

All four data sources provide input to reporting applications. The first three sources are also used as input for these data structures:

- cubes, which are created with either SAS Data Integration Studio or SAS OLAP Cube Studio

- data marts and data warehouses, which are created with SAS Data Integration Studio

You can use these storage options in any combination to meet your unique business requirements.

SAS applications can also access data that is stored in Hadoop file systems. In addition, SAS In-Database enables high-performance computing by running aggregations and analytics inside the database.

The following sections describe each storage option in more detail. Central management of data sources through the SAS metadata repository is also discussed.

Default SAS Storage

You can use SAS data sets (tables), the default SAS storage format, to store data of any granularity.

A SAS table is a file that SAS software creates and processes. Each SAS table is a member of a SAS library. A SAS library is a collection of one or more SAS files that are recognized by SAS software and that are referenced and stored as a unit.

Each SAS table contains the following:

- data values that are organized as a table of observations (rows) and variables (columns) that can be processed by SAS software

- descriptor information such as data types, column lengths, and the SAS engine that was used to create the data

For shared access to SAS tables, you can use SAS/SHARE software, which provides concurrent Update access to SAS files for multiple users.

Third-Party Data Storage

Data can be stored in a wide range of third-party databases, including the following:

- relational databases such as Oracle, Sybase, DB2, SQL Server, and Teradata

- Hadoop data, which is divided into blocks and stored across multiple connected nodes that work together

- hierarchical databases such as IBM Information Management System (IMS)

- Computer Associates Integrated Database Management System (CA-IDMS), which is a network model database system

SAS/ACCESS interfaces provide fast, efficient loading of data to and from these facilities. With these interfaces, SAS software can work directly from the data sources without making a copy. Several of the SAS/ACCESS engines use an input/output (I/O) subsystem that enables applications to read entire blocks of data instead of reading just one record at a time. This feature reduces I/O bottlenecks and enables procedures to read data as fast as they can process it. The SAS/ACCESS engines for Oracle, Sybase, DB2 (on UNIX and PC), ODBC, SQL Server, and Teradata support this functionality. These engines, as well as the DB2 engine on z/OS, can also access database management system (DBMS) data in parallel by using multiple threads to the parallel DBMS server. Coupling the threaded SAS procedures with these SAS/ACCESS engines provides even greater gains in performance.

SAS In-Database enables high-performance computing for complex, high-volume analytics. This technology enables certain Base SAS and SAS/STAT procedures to run aggregations and analytics inside the database. In-database technology minimizes the movement of data across the network, while enabling more sophisticated queries and producing results more quickly. This technology is available through the use of the SAS/ACCESS, SAS Scoring Accelerator, and SAS Analytics Accelerator products. The supported databases include the following:

- Aster

- DB2 (UNIX only)

- Greenplum

- Hadoop

- Netezza

- Oracle

- Teradata

Hadoop Data Storage

Data can be stored as Hadoop data, which is divided into blocks and stored across multiple connected nodes that work together. The benefits of storing data in Hadoop include the following:

- Hadoop accomplishes two tasks: massive data storage and distributed processing.

- Hadoop is a low-cost alternative for data storage over traditional data storage options. Hadoop uses commodity hardware to reliably store large quantities of data.

- Data and application processing are protected against hardware failure. If a node goes down, data is not lost because a minimum of three instances of the data exist in the Hadoop cluster. Furthermore, jobs are automatically redirected to working machines in the cluster.

- The distributed Hadoop model is designed to easily and economically scale up from single servers to thousands of nodes, each offering local computation and storage.

- Unlike traditional relational databases, Hadoop does not require preprocessing of data before storing it. You can easily store unstructured data.

You can use Hadoop to stage large amounts of raw data for subsequent loading into an enterprise data warehouse or to create an analytical store for high-value activities such as advanced analytics, querying, and reporting. SAS enables you to use data stored in Hadoop to do the following:

- explore data and develop and execute models using the following software: SAS Visual Analytics, SAS Visual Statistics, SAS High-Performance Analytics products, SAS In-Database Technology, SAS In-Memory Statistics, and SAS Scoring Accelerator for Hadoop

- access and manage data using the following software: SAS Data Loader for Hadoop, SAS Data Quality Accelerator for Hadoop, SAS In-Database Code Accelerator for Hadoop, SAS/ACCESS interfaces, the Base SAS FILENAME statement and HADOOP procedure, the SAS Scalable Performance Data (SPD) Engine, and SAS Data Integration Studio

- use the features of SAS Event Stream Processing, SAS Federation Server, SAS Grid Manager for Hadoop, SAS High-Performance Marketing Optimization, and SAS Visual Scenario Designer

For more information, see http://support.sas.com/resources/thirdpartysupport/v94/hadoop/.

Parallel Storage

Parallel Storage Overview

The SAS Scalable Performance Data Engine (SPD Engine) and the SAS Scalable Performance Data Server (SPD Server) are designed for high-performance data delivery. They enable rapid access to SAS data for intensive processing by the application.

Although the Base SAS engine is sufficient for most tables that do not span volumes, the SAS SPD Engine and SAS SPD Server are high-speed alternatives for processing very large tables. They read and write tables that contain millions of observations, including tables that expand beyond the 2-GB size limit imposed by some operating systems. In addition, they support SAS analytic software and procedures that require fast processing of tables.

Options for Implementing Parallel Storage

Two options are available for implementing parallel storage:

- The SAS SPD Engine is included with Base SAS software. It is a single-user data storage solution that shares the high-performance parallel processing and parallel I/O capabilities of SAS SPD Server, but lacks the additional complexity of a multi-user server.

  The SPD Engine runs in UNIX and Windows operating environments as well as some z/OS operating environments.

- The SAS SPD Server is available as a separate product. It is a multi-user parallel-processing data server with a comprehensive security infrastructure, backup and restore utilities, and sophisticated administrative and tuning options.

  The SAS SPD Server runs in Windows and UNIX operating environments.

How Parallel Storage Works

The SAS SPD Engine and SAS SPD Server deliver data to applications rapidly by organizing large SAS data sets into a streamlined file format. The file format enables multiple CPUs and I/O channels to perform parallel input/output (I/O) functions on the data. Parallel I/O takes advantage of multiple CPUs and multiple controllers, with multiple disks per controller, to read or write data in independent threads.

One way to take advantage of the features of the SAS SPD Engine is through a hardware and software architecture known as symmetric multiprocessing (SMP). An SMP machine has multiple CPUs and an operating system that supports threads. These machines are usually configured with multiple controllers and multiple disk drives per controller.

When the SAS SPD Engine reads a data file, it launches one or more threads for each of the CPUs in the SMP machine. These threads then read data in parallel from multiple disk drives, driven by one or more controllers per CPU. The SAS SPD Engine running on an SMP machine provides the capability to read and deliver much more data to an application in a given elapsed time.

For more information, see the SAS SPD Server documentation, which is available at http://support.sas.com/documentation/onlinedoc/spds/.

Multidimensional Databases (Cubes)

Multidimensional databases (cubes) are another storage option provided by the SAS Intelligence Platform. Cubes provide business users with multiple views of their data through drill-down capabilities.

Cubes are derived from source data such as SAS tables, SAS SPD Engine tables, and SAS/ACCESS database tables. To create cube definitions, and to build cubes based on these definitions, you can use the Cube Designer wizard, which is available from SAS Data Integration Studio and SAS OLAP Cube Studio.

Cubes are managed by the SAS OLAP Server, which is a multi-user, scalable server designed to store and access large volumes of data while maintaining system performance.

The SAS OLAP Server uses a SAS engine that organizes data into a streamlined file format. This file format enables the engine to rapidly deliver data to client applications. The engine also reads and writes partitioned tables, which enables it to use multiple CPUs to perform parallel I/O functions. The threaded model enables the SAS OLAP Server to create and query aggregations in parallel for fastest performance.

SAS business intelligence applications perform queries against the cubes by using the multidimensional expression (MDX) query language. Cubes can be accessed by client applications that are connected to the SAS OLAP Server with the following tools:

- the SQL pass-through facility for OLAP, which is designed to process MDX queries within the PROC SQL environment

- open access technologies such as OLE DB for OLAP and ADO MD

How Data Sources Are Managed in the Metadata

Creation of Metadata Objects

All of the data sources that are used in your deployment of the SAS Intelligence Platform are centrally controlled through metadata that is stored in the SAS Metadata Repository. In the metadata repository, you can create the following types of metadata objects to control and manage your data:

- database servers, which provide relational database services to clients

- SAS Application Servers, which perform SAS processes on data

- cubes

- OLAP schemas, which specify which groups of cubes a given SAS OLAP Server can access

- dimensions and measures in a cube

- libraries, which are collections of one or more files that are recognized by SAS software and that are referenced and stored as a unit

- the data sources (for example, SAS tables) that are contained in a library

- the columns that are contained in a data source

A variety of methods are available to populate the metadata repository with these objects, including the following:

- The data source design applications, SAS Data Integration Studio and SAS OLAP Cube Studio, automate the creation of all of the necessary metadata about your data sources. As you use these products to define warehouses, data marts, and cubes, the appropriate metadata objects are automatically created and stored in the metadata repository.

- You can use the following features of SAS Management Console to define data source objects:

  - The New Server Wizard enables you to easily define the metadata for your database servers and SAS Application Servers.

  - The Data Library Manager enables you to define database schemas for a wide variety of schema types. You can also use this feature to define libraries if you are not using SAS Data Integration Studio to define them.

  - The Register Tables feature enables you to import table definitions from external sources if you are not using SAS Data Integration Studio to create them.

  - On the **SAS Folders** tab, you can set permissions that secure access to metadata folders and objects. Because all SAS Intelligence Platform applications use the metadata server when accessing resources, permissions that are enforced by the metadata server provide an effective level of protection. These permissions supplement protections from the host environment and other systems. Therefore, a user's ability to perform a particular action is determined not only by metadata-based access controls but also by external authorization mechanisms such as operating system permissions and database controls.

- You can use the metadata LIBNAME engine to enforce data-related Read, Write, Create, and Delete access controls that have been defined in metadata.

For detailed information about administering data sources, see the *SAS Intelligence Platform: Data Administration Guide*.

Information Maps

After your data sources have been defined in metadata, you can use SAS Information Map Studio to create SAS Information Maps, which are business metadata about your data. Information maps enable you to surface your data in business terms that typical business users understand, while storing key information that is needed to build appropriate queries.

SAS Business Data Network

You can use SAS Business Data Network to manage business terms. It supports a collaborative approach to managing the following information:

- descriptions of business terms, including their requirements and attributes

- related source data and reference data

- contacts (such as technical owners, business owners, and interested parties)

- relationships between terms and processes (such as SAS Data Management Studio jobs, services, and business rules)

By linking terms to business rules and data monitoring processes, SAS Business Data Network provides a single entry point for all data consumers to better understand their data. Data stewards, IT staff, and enterprise architects can use the terms to promote a common vocabulary across projects and business units. Collaborative creation, editing, and deletion of terms is enabled through integration with SAS Workflow Studio. For more information, see the *SAS Business Data Network: User's Guide* and the *SAS Workflow Studio: User's Guide*.

SAS Lineage

SAS Lineage is a web-based diagram component for visualizing relationships between objects. It is used as a stand-alone lineage and relationship viewer that can be accessed by SAS data management and business intelligence applications. The component has two modes:

- A network diagram displays all relationships.

- Two left-to-right dependency diagrams are available: one that displays data governance information (governance) and another that displays parent and child relationships only (impact analysis).

For more information, see the *SAS Lineage: User's Guide*.

The relationship information that is displayed in these diagrams is drawn from the Relationship database that is a part of the Web Infrastructure Platform Data Server.

SAS Lineage can display most types of SAS metadata. This includes data objects such as columns, tables, external files, information maps, reports, stored processes, SAS Enterprise Guide projects and associated objects, and the levels and measures in OLAP cubes. You can also display objects created in SAS Business Data Network, such as terms, tags, and associated items. In addition, lineage information from third-party applications can be imported to the relationship database using SAS Metadata Bridges.

Chapter 4
Servers in the SAS Intelligence Platform

Overview of SAS Servers

The SAS Intelligence Platform provides access to SAS functionality through the following specialized servers:

Note: In the SAS Intelligence Platform, the term "server" refers to a program or programs that wait for and fulfill requests from client programs for data or services. The term server does not necessarily refer to a specific computer, since a single computer can host one or more servers of various types.

- the SAS Metadata Server, which writes metadata objects to, and reads metadata objects from, SAS Metadata Repositories. These metadata objects contain information about all of the components of your system, such as users, groups, data libraries, servers, and user-created products such as reports, cubes, and information maps.

- SAS Workspace Servers, which provide access to SAS software features such as the SAS language, SAS libraries, the server file system, results content, and formatting services.

  A program called the SAS object spawner runs on a workspace server's host machine. The spawner listens for incoming client requests and launches server instances as needed.

- SAS Pooled Workspace Servers, which are configured to use server-side pooling. In this configuration, the SAS object spawner maintains a collection of workspace server processes that are available for clients. This server configuration is intended for use by query and reporting tools such as SAS Information Map Studio, SAS Web Report Studio, and the SAS Information Delivery Portal.

- SAS Stored Process Servers, which fulfill client requests to execute SAS Stored Processes. Stored processes are SAS programs that are stored on a server and can be executed as required by requesting applications. By default, three load balanced SAS Stored Process Servers are configured.

  The SAS object spawner runs on a stored process server's host machine, listens for incoming client requests, and launches server instances as needed.

- SAS OLAP Servers, which provide access to cubes. Cubes are logical sets of data that are organized and structured in a hierarchical multidimensional arrangement. Cubes are queried by using the multidimensional expression (MDX) language.

- the SAS Web Infrastructure Platform Data Server, which is the default location for middle-tier data such as alerts, comments, and workflows, as well as data for the SAS Content Server. The server, which is backed by PostgreSQL, is provided as an alternative to using a third-party DBMS. (The server cannot be used as a general-purpose data store.)

- SAS/CONNECT servers, which provide computing resources on remote machines where SAS Integration Technologies is not installed.

- batch servers, which enable you to execute code in batch mode. There are three types of batch servers: DATA step batch servers, Java batch servers, and generic batch servers. The DATA step server enables you to run SAS DATA steps and procedures in batch mode. The Java server enables you to schedule the execution of Java code, such as the code that creates a SAS Marketing Automation marketing campaign. The generic server supports the execution of any other type of code.

- the SAS Content Server, which is part of the middle tier architecture. This server stores digital content (such as documents, reports, and images) that is created and used by SAS web applications. For more information, see "SAS Content Server " on page 39.

- the SAS LASR Analytic Server, which is the in-memory analytics platform for SAS Visual Analytics, SAS Visual Statistics, and other high-performance products.

Note: For accessing specialized data sources, the SAS Intelligence Platform can also include one or more data servers. These might include the SAS Scalable Performance Data (SPD) Server and third-party database management system (DBMS) products. The SAS OLAP Server also provides some data server functionality. For information about data servers, see Chapter 3, "Data in the SAS Intelligence Platform," on page 17.

SAS Metadata Server

About the SAS Metadata Server

The SAS Metadata Server is a multi-user server that enables users to read metadata from and write metadata to one or more SAS Metadata Repositories. This server is a centralized resource for storing, managing, and delivering metadata for all SAS applications across the enterprise.

The SAS Metadata Server can be set up in a clustered configuration to provide redundancy and high availability.

About the Metadata in the SAS Metadata Repository

Here are examples of the types of metadata objects that can be stored in the SAS Metadata Repository:

- users
- groups of users
- data libraries
- tables
- jobs
- cubes
- documents
- information maps
- reports
- stored processes
- dashboards
- analytical models
- SAS Workspace Servers
- SAS Pooled Workspace Servers
- SAS Stored Process Servers
- SAS OLAP Servers

A metadata object is a set of attributes that describe a resource. Here are some examples:

- When a user creates a report in SAS Web Report Studio, a metadata object is created to describe the new report.
- When a user creates a data warehouse in SAS Data Integration Studio, a metadata object is created to describe each table included in the warehouse.
- When a system administrator defines a new instance of a SAS server, a metadata object is created to describe the server.

The specific attributes that a metadata object includes vary depending on the resource that is being described. For example, a table object can include attributes for the table's name and description, path specification, host type, and associated SAS Application Server.

The SAS Metadata Server uses the SAS Open Metadata Architecture, which provide common metadata services to SAS and other applications. Third parties can access metadata in the SAS Metadata Server by using an application programming interface (API) that is supplied by SAS. SAS Metadata Bridges are available to support metadata exchange with a variety of sources, including the Common Warehouse Metadata Model (CWM).

How the Metadata Server Controls System Access

The SAS Metadata Server plays an important role in the security of the SAS Intelligence Platform. It controls system access in the following ways:

SAS identities
> For accountability, we recommend that you create an individual SAS identity for each person who uses the SAS environment. These identities enable you to make access distinctions and audit individual actions in the metadata layer. The identities also provide personal folders for each user. The metadata server maintains its own copy of each user ID for the purpose of establishing a SAS identity.

access controls
> You can define metadata-based access controls that supplement protections provided by the host environment and other systems. The metadata-based controls enable you to manage access to OLAP data, to relational data (depending on the method by which the data is accessed), and to almost any metadata object (for example, reports, data definitions, information maps, jobs, stored processes, and server definitions).

roles
> You can assign users and groups to roles that determine whether they can use application features such as menu items, plug-ins, and buttons. Roles are designed to make application functionality available to the appropriate types of users. For example, role memberships determine whether a user can see the Server Manager plug-in (in SAS Management Console), compare data (in SAS Enterprise Guide), or directly open an information map (in SAS Web Report Studio). Applications that support roles include the SAS Add-In for Microsoft Office, SAS Enterprise Guide, SAS Forecast Studio, SAS Management Console, and SAS Web Report Studio.

For more information about security in the SAS Intelligence Platform, see Chapter 7, "Security Overview," on page 53.

How Metadata Is Created and Administered

Metadata can be loaded to the SAS Metadata Server in several ways:

- The configuration process for the SAS Intelligence Platform automatically creates and stores metadata objects for the resources, such as servers, that are part of your initial installation.

- SAS Metadata Bridges enable you to import metadata from a variety of sources, including market-leading design tool and repository vendors and the Common Warehouse Metamodel (CWM).

- When users create content such as reports, information maps, and data warehouses with the SAS Intelligence Platform applications, these applications create and store metadata objects describing the content.

- System administrators use the SAS Management Console client application to directly create metadata for system resources such as servers, users, and user groups.

System administrators also use SAS Management Console for general metadata administration tasks, including backing up the metadata server, creating new repositories, promoting metadata objects, and maintaining authorization information and access rules for all types of resources.

How Business Metadata Is Organized

A hierarchical folder structure is used to organize metadata for business intelligence content such as libraries, tables, jobs, cubes, information maps, and reports. The folder structure includes personal folders for individual users and a folder for shared data. Within this overall structure, you can create a customized folder structure that meets the information management, data sharing, and security requirements of your organization.

SAS Management Console and SAS Environment Manager display all SAS folders that the user has permission to view. Most other client applications display SAS folders only if they contain content that is relevant to the application, subject to the user's permissions. Import and export wizards in SAS Management Console, SAS Data Integration Studio, and SAS OLAP Cube Studio enable you to easily move or promote content from one part of the SAS Folders tree to another, or from a test environment to a production environment.

For more information about the SAS Metadata Server, see the *SAS Intelligence Platform: System Administration Guide*.

Server Objects, Application Servers, and Logical Servers

About Server Objects and Server Groupings

In the SAS Metadata Repository, each server process that executes SAS code is represented by a server object. In the metadata, the attributes for each server object contain information such as the following:

- the name of the machine that is hosting the server
- the TCP/IP port or ports on which the server listens for requests
- the SAS command that is used to start the server

The intermediate level of organization is called a logical server object. SAS servers of a particular type, running either on the same machine or on different machines, can be grouped into a logical server of the corresponding type. For example:

- A logical workspace server is a group of one or more workspace servers.
- A logical pooled workspace server is a group of one or more workspace servers that are configured to use server-side pooling.
- A logical stored process server is a group of one or more stored process servers.

The logical servers are then grouped into a SAS Application Server. The following figure shows a sample configuration:

Figure 4.1 *SAS Application Server Components: Sample Configuration*

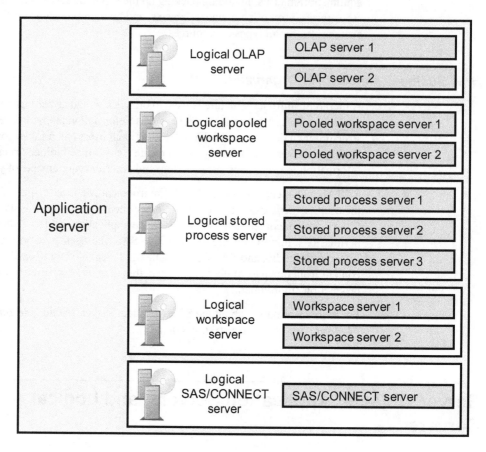

Application servers and logical servers are logical constructs that exist only in metadata. In contrast, the server objects within a logical server correspond to actual server processes that execute SAS code.

Purpose of the SAS Application Server Grouping

SAS application servers, which are groupings of logical servers, provide the following functionality in the SAS Intelligence Platform:

- SAS applications can request the services of a group of logical servers by specifying a single application server name.

  For example, a user can set up SAS Data Integration Studio to send all jobs to a specific SAS Application Server. Each time the user runs a SAS Data Integration Studio job, the SAS code that is generated is executed by a workspace server that is contained in the specified application server.

- System resources, such as SAS libraries or data schemas, can be assigned to a specific application server. This assignment ensures that all of the logical servers contained in the application server can access these resources as needed, subject to security restrictions.

  For example, if a SAS library is assigned to a specific application server, then any application that runs jobs on that server will automatically have access to the library, subject to security restrictions.

Purpose of the Logical Server Grouping

Logical servers, which are groupings of individual servers of a specific type, provide the following functionality in the SAS Intelligence Platform:

- Users or groups of users can be granted rights to a logical server, thereby providing access to each physical server that the logical server contains. You can also use permissions to direct specific users or groups of users to specific physical servers within a logical server grouping.

- If one physical server in a logical server grouping fails, the other servers are still available to provide continuous processing.

- Load balancing can be implemented among the servers in a logical workspace server, logical stored process server, or logical OLAP server. Load balancing is configured by default for the SAS Pooled Workspace Server and SAS Stored Process Servers.

- Workspace pooling can be implemented among the servers in a logical workspace server.

For more information about SAS Application Servers, see the *SAS Intelligence Platform: Application Server Administration Guide*.

Load Balancing for SAS Workspace Servers, SAS Stored Process Servers, and SAS OLAP Servers

Load balancing is a feature that distributes work among the server processes in a logical workspace server, logical stored process server, or logical OLAP server. The load balancer runs in the object spawner, which is a program that runs on server machines, listens for incoming client requests, and launches server instances as needed. When a logical server is set up for load balancing, and the object spawner receives a client request for a server in the logical server group, the spawner directs the request to the server in the group that has the least load.

SAS Pooled Workspace Servers and SAS Stored Process Servers are load-balanced by default. For more information about load balancing, see the *SAS Intelligence Platform: Application Server Administration Guide*.

Workspace Pooling for SAS Workspace Servers

Workspace pooling creates a set of server connections that are reused. This reuse avoids the wait times that an application incurs when it creates a new server connection for each user. You can also use pooling to distribute server connections across machines.

Pooling is most useful for applications that require frequent, but brief, connections to a SAS Workspace Server. Two types of pooling are supported:

server-side pooling
: a configuration in which the SAS object spawner maintains a collection of re-usable workspace server processes that are available for clients. The usage of servers in the pool is governed by authorization rules that are set on the servers in the SAS metadata. Load balancing is automatically configured for these servers.

By default, applications such as SAS Information Map Studio, the SAS Information Delivery Portal, and SAS Web Report Studio use SAS Pooled Workspace Servers to query relational data. SAS Pooled Workspace Servers are configured to use server-side pooling and load balancing.

client-side pooling

a configuration in which the client application maintains a collection of reusable workspace server processes. In releases prior to 9.2, client-side pooling was the only method of configuring pooling for workspace servers.

For more information about pooling, see the *SAS Intelligence Platform: Application Server Administration Guide.*

Logging and Monitoring for SAS Servers

The SAS Intelligence Platform uses a standard logging facility to perform logging for SAS servers. The logging facility supports problem diagnosis and resolution, performance and capacity management, and auditing and regulatory compliance. The logging facility provides the following capabilities for servers:

- Log events can be directed to multiple destinations, including files, operating system facilities, and client applications.

- For each log destination, you can configure the message layout, including the contents, the format, the order of information, and literal text.

- For each log destination, you can configure a filter to include or exclude events based on levels and message contents.

- For the metadata server, security-related events are captured, including authentication events, client connections, changes to user and group information, and permission changes.

- You can generate performance-related log events in a format that can be processed by an Application Response Measurement (ARM) 4.0 server.

SAS Environment Manager provides web-based management, operation, and proactive monitoring of servers on both the middle tier and the SAS server tier. Backed by Hyperic technology from VMware, this application provides automatic resource discovery, monitoring of remote systems, role-based dashboards, alerting, and visualization features.

In addition, you can use the server management features of SAS Management Console to do the following:

- view server and spawner logs

- change logging levels dynamically without stopping the server

- quiesce, stop, pause, and resume servers and spawners

- validate servers and spawners and test server connections

- view information about current connections and processes

- view performance counters that provide statistics about activity that has occurred since a server or spawner was last started

Server monitoring can also be performed using third-party products for enterprise systems management.

For more information about server logging and monitoring, see the *SAS Intelligence Platform: System Administration Guide*, the *SAS Logging: Configuration and Programming Reference*, and the SAS Environment Manager Help.

SAS Grid Computing

You can use SAS grid computing tools to manage a distributed grid environment for your SAS deployment. SAS Grid Manager, working together with Platform Suite for SAS, enables you to distribute server workloads across multiple computers on a network to obtain the following benefits:

- the ability to accelerate SAS analytical results by adding additional computers to the grid and by dividing jobs into separate processes that run in parallel across multiple servers

- the flexibility to upgrade and maintain the computing resources on which your SAS servers are deployed without disrupting operations, and to add computing resources quickly to handle increased workloads and peak demands

- continuity of service through the high availability of critical components running in the grid

Implementation of a grid environment involves planning and design efforts to determine the topology and configuration that best meets the needs of your organization. In some cases, third-party data sharing facilities or hardware load balancers might be required. For more information, see *Grid Computing in SAS* and "Introduction to Grid Computing" at http://support.sas.com/rnd/scalability/grid/index.html.

SAS LASR Analytic Server

The SAS LASR Analytic Server is an analytic platform that is provided with SAS Visual Analytics, SAS Visual Statistics, and other high-performance products. This secure, multi-user server provides concurrent access to data that is loaded into memory. The server can take advantage of a distributed computing environment by distributing data and the workload among multiple machines and performing massively parallel processing. The server can also be deployed on a single machine where the workload and data volumes do not demand a distributed computing environment but can still benefit from the speed of in-memory processing.

The server handles both big data and smaller sets of data, and it is designed with high-performance, multi-threaded, analytic code. The server processes client requests at extraordinarily high speeds due to the combination of hardware and software that is designed for rapid access to tables in memory. By loading tables into memory for analytic processing, the server enables business analysts to explore data and discover relationships in data at the speed of the RAM that is installed on the system.

The server can also perform text analysis on unstructured data. The unstructured data is loaded to memory in the form of a table, with one document in each row.

Data can be loaded into a distributed server in the following ways:

- You can load tables into the server by using the SAS LASR Analytic Server engine or the LASR procedure from a SAS session that has a network connection to the cluster. Any data source that can be accessed with a SAS engine can be loaded into memory. The data is transferred to the root node of the server, and the root node distributes the data to the worker nodes. You can also append rows to an in-memory table with the SAS LASR Analytic Server engine.

- Tables can be read from the Hadoop Distributed File System (HDFS) or an NFS-mounted distributed file system. You can use the SASHDAT engine to add tables to HDFS. When a table is added to HDFS, it is divided into blocks that are distributed across the machines in the cluster. The server software is designed to read data in parallel from HDFS. When used to read data from HDFS, the LASR procedure causes the worker nodes to read the blocks of data that are local to the machine.

- Tables can be read from a third-party vendor database. For distributed databases like Teradata and Greenplum, the SAS LASR Analytic Server can access the data in the appliance.

- Tables can be loaded using SAS Visual Analytics Administrator.

For more information, see *SAS LASR Analytic Server: Reference Guide*.

Chapter 5
Middle-Tier Components of the SAS Intelligence Platform

Overview of Middle-Tier Components

The middle tier of the SAS Intelligence Platform provides an environment in which the business intelligence web applications, such as SAS Web Report Studio and the SAS Information Delivery Portal, can execute. These products are initiated from a browser interface. They run in a web application server and communicate with the user by sending data to and receiving data from the user's web browser. The middle tier applications rely on servers on the SAS server tier to perform SAS processing, including data query and analysis.

The following figure illustrates the middle-tier components:

Figure 5.1 Middle Tier Components

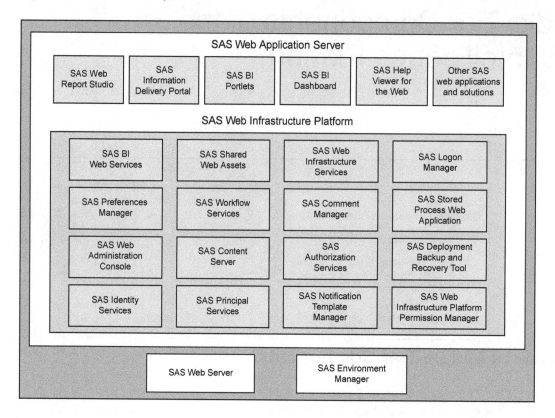

SAS Middle Tier Servers

The SAS Web Server and SAS Web Application Server, which are included with the SAS Intelligence Platform, provide a highly scalable, easy-to-manage environment that is dedicated to running SAS web applications. The SAS Web Server provides static HTTP content to users. It forwards requests for dynamic content to the SAS Web Application Server, which provides an execution environment for the following applications and services:

• web applications including SAS Web Report Studio, SAS Information Delivery Portal, SAS BI Portlets, and SAS BI Dashboard. For descriptions, see Chapter 6, "Clients in the SAS Intelligence Platform," on page 41. (Depending on which products and solutions you have licensed, your site might have additional web applications.)

• applications and services that are part of the SAS Web Infrastructure Platform.

To meet the processing demands of your organization, multiple SAS Web Application Server instances can be deployed in a clustered configuration. Based on your specifications, server configuration is performed automatically during the deployment process. The servers use the Java Runtime Environment (JRE) software, which is provided as part of the SAS installation. No middle tier software needs to be obtained from third parties.

SAS Environment Manager provides a comprehensive interface for ongoing management, operation, and proactive monitoring of servers on both the middle tier and the SAS server tier.

SAS Web Infrastructure Platform

The SAS Web Infrastructure Platform is a collection of services and applications that provide common infrastructure and integration features to be used by SAS web applications. These services and applications provide the following benefits:

- consistency in installation, configuration, and administration tasks for web applications

- greater consistency in users' interactions with web applications

- integration among web applications as a result of the ability to share common resources

The following services and applications are included in the SAS Web Infrastructure Platform:

SAS Authorization Service
  is used by some SAS web applications that manage authorization through web services.

SAS BI Web Services for Java
  can be used to enable your custom applications to invoke and obtain metadata about SAS Stored Processes. Web services enable distributed applications that are written in different programming languages and that run on different operating systems to communicate using standard web-based methods, including REST and SOAP.

SAS Content Server
  stores digital content (such as documents, reports, and images) that is created and used by SAS web applications. For more information, see "SAS Content Server " on page 39.

SAS Deployment Backup and Recovery Tool
  enables deployment-wide backup and recovery services.

SAS Identity Services
  provides SAS web applications with access to user identity information.

SAS Logon Manager
  provides a common user authentication mechanism for SAS web applications. It displays a dialog box for user ID and password entry, authenticates the user, and launches the requested application. SAS Logon Manager supports a single sign-on authentication model. When this model is enabled, it provides access to a variety of computing resources (including servers and web pages) during the application session without repeatedly prompting the user for credentials.

  You can configure SAS Logon Manager to display custom messages and to specify whether a logon dialog box is displayed when users log off. In addition, you can use third-party products in conjunction with SAS Logon Manager to enable users to access multiple web applications with a single sign-on action.

SAS Notification Template Editor
  enables administrators to create and edit messages that are sent as notifications to end users of SAS applications.

SAS Preferences Manager

> provides a common mechanism for managing preferences for SAS web applications. The feature enables administrators to set default preferences for locale, theme, alert notification, and time, date, and currency display. Within each web application, users can view the default settings and update their individual preferences.

SAS Principal Services

> enables access to core platform web services for SAS applications.

SAS Shared Web Assets

> contains graph applet JAR files that are shared across SAS web applications. They display graphs in stored processes and in the SAS Stored Process web application.

SAS Stored Process Web Application

> executes stored processes on behalf of a web client and return results to a web browser. The SAS Stored Process web application is similar to the SAS/IntrNet Application Broker, and has the same general syntax and debug options. Web applications can be implemented using the SAS Stored Process Web Application, the Stored Process Service API, or a combination of both.

SAS Web Administration Console

> provides features for monitoring and administering middle-tier components. This browser-based interface enables administrators to do the following:

> - monitor information about users who are currently logged on to SAS web applications
>
> - view audit reports of logon and logoff activity
>
> - manage notification templates and letterheads
>
> - manage authorization, including Web Infrastructure Platform roles and privileges and web-layer permissions
>
> - view configuration information for each middle-tier component

SAS Web Infrastructure Platform Permission Manager

> enables administrators to set web-layer permissions on folders and documents for SAS applications that use SAS Content Services for access to digital content. You can access the permissions manager with the SAS Web Administration Console.

SAS Web Infrastructure Platform Services

> provides a common infrastructure for SAS web applications. The infrastructure supports activities such as auditing, authentication, configuration, status and monitoring, email, theme management, alert registration and notification, collaboration, and data sharing across SAS web applications.

SAS Workflow Services

> enables the deployment and execution of workflows that are integrated with certain SAS applications and solutions. The SAS Workflow Studio client application enables business users to quickly create or modify workflow templates for execution by workflow services.

By default, the SAS Web Infrastructure Platform uses a relational database on the SAS Web Infrastructure Platform Data Server to store alerts, comments, workflows, and SAS Content Server data. During installation, you can choose to use a third-party vendor database instead of the SAS Web Infrastructure Platform Data Server.

For more information about the SAS Web Infrastructure Platform, see the *SAS Intelligence Platform: Middle-Tier Administration Guide.*

SAS Content Server

The SAS Content Server is part of the SAS Web Infrastructure Platform. This server stores digital content (such as documents, reports, and images) that is created and used by SAS web applications. For example, the SAS Content Server stores report definitions that are created by users of SAS Web Report Studio, as well as images and other elements that are used in reports. A process called content mapping ensures that report content is stored using the same folder names, folder hierarchy, and permissions that the SAS Metadata Server uses to store corresponding report metadata.

In addition, the SAS Content Server stores documents and other files that are to be displayed in the SAS Information Delivery Portal or in SAS solutions.

To interact with the SAS Content Server, client applications use WebDAV-based protocols for access, versioning, collaboration, security, and searching. Administrative users can use the browser-based SAS Web Management Console to create, delete, and manage permissions for folders on the SAS Content Server.

For more information about the SAS Content Server, see the *SAS Intelligence Platform: Middle-Tier Administration Guide*.

Chapter 6
Clients in the SAS Intelligence Platform

Overview of Clients

SAS Intelligence Platform clients include desktop applications and web-based applications. The following table lists the clients by type:

Table 6.1 *SAS Intelligence Platform Clients*

Desktop Applications	Web Applications
Java Applications: • SAS Data Integration Studio • SAS Enterprise Miner • SAS Forecast Studio • SAS Information Map Studio • SAS Management Console • SAS Model Manager • SAS OLAP Cube Studio • SAS Workflow Studio Windows applications:* • SAS Add-In for Microsoft Office • SAS Enterprise Guide • JMP	• SAS BI Portlets • SAS BI Dashboard • SAS Environment Manager • SAS Help Viewer for the Web • SAS Information Delivery Portal • SAS Visual Analytics • SAS Visual Statistics • SAS Web Parts for Microsoft SharePoint • SAS Web Report Studio

* JMP is also available on Macintosh and Linux.

The desktop applications are supported only on Microsoft Windows systems. The exception is SAS Management Console, which runs on all platforms except z/OS. All of the Java desktop applications use the Java Runtime Environment (JRE), provided with SAS, which executes the application and a set of standard Java class libraries.

The web-based applications reside and execute on the middle tier. They require only a web browser to be installed on each client machine, with the addition of Adobe Flash Player for SAS BI Dashboard. The applications run on the SAS Web Application Server and communicate with the user by sending data to and receiving data from the user's browser. For example, an application of this type displays its user interface by sending an HTML document to the user's browser. The user can submit input to the application by sending it an HTTP response—usually by clicking a link or submitting an HTML form.

To supplement these clients, mobile devices (including iPad and Android devices) can be used to view certain types of reports. This capability is enabled through SAS Mobile BI.

SAS Add-In for Microsoft Office

The SAS Add-In for Microsoft Office enables you to harness the power of SAS analytics from Microsoft Word, Microsoft Excel, Microsoft PowerPoint, and Microsoft Outlook. You can use the SAS Add-In for Microsoft Office for the following tasks:

• generate basic statistics and charts for a data source using the Quick Start tools.

• analyze SAS or Excel data using SAS tasks.

• embed SAS content in your documents, spreadsheets, and presentations.

• access and view SAS data sources or any data source that is available from your SAS server. There is no size limit on the SAS data sources that you can open.

- edit a SAS data source in Microsoft Excel and save your changes to the server.
- refresh the SAS content to display the most updated results. The results could change because of changes to the underlying data or changes to the SAS task or stored process that generates the results.
- view and monitor SAS reports in Microsoft Outlook. You can share these reports by sending email, scheduling meetings, or assigning tasks in Microsoft Outlook. You can also share a report using an instant messaging tool or by sending the report to another Microsoft Office application, such as Microsoft Excel, Microsoft Word, or Microsoft PowerPoint.

For more information, see the Help for the SAS Add-In for Microsoft Office. For information about administrative tasks associated with SAS Add-In for Microsoft Office, see the *SAS Intelligence Platform: Desktop Application Administration Guide*.

SAS BI Dashboard

The SAS BI Dashboard enables users to create, maintain, and view dashboards to monitor key performance indicators that convey how well an organization is performing. The application is web-based, leveraging the capabilities of Adobe Flash Player, and can be accessed from within the SAS Information Delivery Portal.

The SAS BI Dashboard includes an easy-to-use interface for creating dashboards that include graphics, text, colors, and hyperlinks. Dashboards can link to SAS reports and analytical results, SAS Strategic Performance Management scorecards and objects, externally generated data, and virtually anything that is addressable by a Uniform Resource Identifier (URI).

All content is displayed in a role-based, secure, customizable, and extensible environment. End users can customize how information appears on their personal dashboards.

For more information, see the SAS BI Dashboard Help, which is available from within the product, and the *SAS BI Dashboard: User's Guide*, available at http://support.sas.com. For information about administrative tasks associated with the SAS BI Dashboard, see the *SAS Intelligence Platform: Web Application Administration Guide*.

SAS BI Portlets

SAS BI Portlets enable users to access, view, and work with business intelligence content that is stored on the SAS Metadata Server and the SAS Content Server. The portlets are seamlessly integrated into the SAS Information Delivery Portal and provide added value to portal users. The portlets include the SAS Collection Portlet, the SAS Diagnostics Portlet, the SAS Navigator Portlet, the SAS Report Portlet, and the SAS Stored Process Portlet. For information about configuring, deploying, and administering SAS BI Portlets, see the *SAS Intelligence Platform: Web Application Administration Guide*.

SAS Data Integration Studio

SAS Data Integration Studio is a visual design tool that enables you to consolidate and manage enterprise data from a variety of source systems, applications, and technologies. This software enables you to create process flows that accomplish the following tasks:

- extract, transform, and load data for use in data warehouses and data marts

- cleanse, migrate, synchronize, replicate, and promote data for applications and business services

SAS Data Integration Studio enables you to create metadata that defines sources, targets, and the processes that connect them. This metadata is stored in one or more shareable repositories. SAS Data Integration Studio uses the metadata to generate or retrieve SAS code that reads sources and creates targets in physical storage. Other applications that share the same repositories can use the metadata to access the targets and use them as the basis for reports, queries, or analyses.

Through its metadata, SAS Data Integration Studio provides a single point of control for managing the following resources:

- data sources (from any platform that is accessible to SAS and from any format that is accessible to SAS)

- data targets (to any platform that is accessible to SAS, and to any format that is supported by SAS)

- processes that specify how data is extracted, transformed, and loaded from a source to a target

- jobs that organize a set of sources, targets, and processes (transformations)

- source code generated by SAS Data Integration Studio

- user-written source code, including legacy SAS programs

For more information, see the SAS Data Integration Studio Help, which is available from within the product, and the *SAS Data Integration Studio: User's Guide*, which is available at http://support.sas.com. For information about administrative tasks associated with SAS Data Integration Studio, see the *SAS Intelligence Platform: Desktop Application Administration Guide*.

SAS Enterprise Guide

SAS Enterprise Guide is a project-oriented application that is designed to enable quick access to much of the analytic power of SAS software for statisticians, business analysts, and SAS programmers. SAS Enterprise Guide provides the following functionality:

- a point-and-click user interface to SAS servers

- transparent data access to both SAS data and other types of data

- interactive task windows that lead you through dozens of analytical and reporting tasks

- a Program Editor with autocomplete and integrated syntax help

- the ability to use the highest quality SAS graphics

- the ability to export results to other Windows applications and the web

- the ability to schedule your project to run at a later time

- OLAP data access, visualization, and manipulation

SAS Enterprise Guide also enables you to create SAS Stored Processes and to store that code in a repository that is available to a SAS Stored Process Server. (Stored processes are SAS programs that are stored on a server and are executed by client applications.) Stored processes are used for web reporting and analytics, among other things.

For more information, see the SAS Enterprise Guide Help, which is available from within the product. For information about administrative tasks associated with SAS Enterprise Guide, see the *SAS Intelligence Platform: Desktop Application Administration Guide.*

SAS Enterprise Miner

SAS Enterprise Miner is an example of the SAS analytics products that use the SAS Intelligence Platform. This application streamlines the data mining process to create highly accurate predictive and descriptive models based on the analysis of vast amounts of data from across an enterprise. SAS Enterprise Miner enables you to develop data mining flows that consist of data exploration, data transformations, and modeling tools to predict or classify business outcomes.

For more information, see the SAS Enterprise Miner Help, which is available from within the product. For information about administrative tasks associated with SAS Enterprise Miner, see the *SAS Intelligence Platform: Desktop Application Administration Guide.*

SAS Environment Manager

SAS Environment Manager is a web-based administration solution for a SAS environment. The application enables you to perform these tasks:

- administer, monitor, and manage SAS resources, including administering the SAS Web Application Server and monitoring SAS foundation servers

- collect and chart data on metrics for monitored SAS resources, thus creating a comprehensive view of resource health and operation

- monitor system and log events

- trigger alerts based on user-specific events and thresholds

- incorporate the monitoring and managing of IT and SAS resources into a service management strategy. This strategy enables you to use the detailed metric information stored in the SAS Environment Manager Data Mart and the reports provided in the Report Center.

- perform SAS administration functions including authorization, user management, server management, library management, and backup.

SAS Environment Manager is based on the VMware Hyperic product, with customizations and plug-ins to optimize the product specifically for a SAS environment. The basic architecture consists of an agent process (running on each middle-tier and server-tier machine) that communicates with a central management server. Agents monitor detected resources and periodically report resource metrics back to the server. The server provides an interface for interacting with those agents. You can use it to manage the data collected by the agents, distribute plug-ins, create alerts and escalation procedures based on collected metrics, and graph the metrics provided through the installed plug-ins.

As of SAS 9.4M5, you can use the SAS Deployment Wizard to configure SAS Environment Manager to use HTTPS. This option offers a simplified route to configuring security. To configure TLS, some post-deployment steps are still required. For more information, see the SAS Intelligence Platform: Middle-Tier Administration Guide (http://documentation.sas.com/? cdcId=bicdc&cdcVersion=9.4&docsetId=bimtag&docsetTarget=p1fpnnm9hxkhlzn1x5tk qs1caeg5.htm&locale=en).

SAS Environment Manager agents run on all SAS platforms except for z/OS. For more information, see *SAS Environment Manager: User's Guide*, *SAS Environment Manager Administration: User's Guide*, and the Help for SAS Environment Manager, which is available from within the product.

SAS Forecast Studio

SAS Forecast Studio is the client component of SAS Forecast Server, which is an example of the SAS analytics products that use the SAS Intelligence Platform. SAS Forecast Server is a large-scale, automatic forecasting solution that enables organizations to produce huge quantities of high-quality forecasts quickly and automatically. SAS Forecast Studio provides a graphical interface (based on Java) to the forecasting and time series analysis procedures contained in SAS High-Performance Forecasting and SAS/ETS software.

SAS High-Performance Forecasting automatically selects the appropriate model for each item being forecast, based on user-defined criteria. Holdout samples can be specified so that models are selected not only by how well they fit past data but by how appropriate they are for predicting the future. If the best forecasting model for each item is unknown or the models are outdated, a maximum level of automation can be chosen in which all three forecasting steps (model selection, parameter estimation, and forecast generation) are performed. If suitable models have been determined, you can keep the current models and reestimate the model parameters and generate forecasts. For maximum processing speed, you can keep previously selected models and parameters and choose to simply generate the forecasts.

For more information, see the *SAS High-Performance Forecasting: User's Guide* and the *SAS/ETS User's Guide*.

SAS Information Delivery Portal

The SAS Information Delivery Portal enables you to aggregate data from a variety of sources and present the data in a web browser. The web browser content might include the output of SAS Stored Processes, links to web addresses, documents, syndicated content from information providers, SAS Information Maps, SAS reports, and web applications.

Using the portal, you can distribute different types of content and applications as appropriate to internal users, external customers, vendors, and partners. You can use the portal along with the Publishing Framework to publish content to SAS publication channels or WebDAV repositories, to subscribe to publication channels, and to view packages published to channels. The portal's personalization features enable users to organize information about their desktops in a way that makes sense to them.

For more information, see the SAS Information Delivery Portal Help, which is available from within the product. For information about administrative and development tasks associated with the SAS Information Delivery Portal, see the *SAS Intelligence Platform: Web Application Administration Guide* and *Developing Portlets for the SAS Information Delivery Portal*, which are available at http://support.sas.com.

SAS Information Map Studio

SAS Information Map Studio enables data modelers and data architects to create and manage SAS Information Maps, which are business metadata about your data. Information maps enable you to surface your data warehouse data in business terms that typical business users understand. It simultaneously while stores key information that is needed to build appropriate queries.

Information maps provide the following benefits:

- Users are shielded from the complexities of the data.

- Data storage is transparent to the consumers of information maps, regardless of whether the underlying data is relational or multidimensional, in a SAS data set, or in a third-party database system.

- Business formulas and calculations are predefined, which makes them usable on a consistent basis.

- Users can easily query data for answers to business questions without having to know query languages.

For more information, see the SAS Information Map Studio Help, which is available from within the product, and *SAS Information Map Studio: Getting Started with SAS Information Maps*, which is available at http://support.sas.com.

JMP

JMP is interactive, exploratory data analysis and modeling software for the desktop. JMP makes data analysis—and the resulting discoveries—visual and helps communicate those discoveries to others. JMP presents results both graphically and numerically. By linking graphs to each other and to the data, JMP makes it easier to see the trends, outliers, and other patterns that are hidden in your data.

Through interactive graphs that link statistics and data, JMP offers analyses from the basic (univariate descriptive statistics, ANOVA, and regression) to the advanced (generalized linear, mixed, and nonlinear models, data mining, and time series models). A unified, visual approach makes these techniques available to all levels of users.

The JMP Scripting Language can be used to create interactive applications and to communicate with SAS. JMP reads text files, Microsoft Excel spreadsheets, SAS data sets, and data from any ODBC data source.

SAS Management Console

SAS Management Console provides a single point of control for administering your SAS servers and for managing metadata objects that are used throughout the SAS Intelligence Platform. You can use SAS Management Console to connect to the SAS Metadata Server and view and manage the metadata objects that are stored in the server's metadata repositories.

SAS Management Console uses an extensible plug-in architecture, enabling you to customize the console to support a wide range of administrative capabilities. The plug-ins that are provided with the application enable you to manage the following resources:

- server definitions

- library definitions

- user, group, and definitions

- resource access controls

- metadata repositories

- job schedules

The plug-ins also enable you to monitor server activity, configure and run metadata server backups, view metadata interdependencies, and stop, pause, and resume servers. Only certain users can view and use plug-ins. A user's access to plug-ins depends on the which roles the user is assigned to and which capabilities are assigned to those roles.

In SAS Management Console, you can also view the folders that store metadata for business intelligence content such as libraries, tables, jobs, cubes, information maps, and reports. The folders include a personal folder for each individual user. For shared data, you can create a customized folder structure and assign appropriate permissions at each level in the folder hierarchy. Import and export wizards enable you to easily move or promote content from one part of the folder tree to another, or from one metadata server to another.

Note: Other client applications also display folders if they contain content that is relevant to the application and that the user has permission to view.

For more information, see the SAS Management Console Help, which is available from within the product. For detailed information about specific tasks that you can perform in SAS Management Console, see the administration guides for the SAS Intelligence Platform at http://support.sas.com/94administration.

SAS Mobile BI

SAS Mobile BI enables users to use mobile devices to view certain relational reports that have been created with SAS Web Report Studio. Each of the supported devices (iPad and Android) displays the reports in its native format.

A graphical user interface enables administrators to manage device eligibility by exclusion or inclusion. SAS metadata security is enforced for all reports.

For more information, see the *SAS Intelligence Platform: Middle-Tier Administration Guide.*

SAS Model Manager

SAS Model Manager provides a central model repository and model management environment for predictive and analytical models. Users can organize modeling projects, develop and validate candidate models, assess candidate models for champion model and challenger model selection, publish and monitor champion models in a production environment, and retrain models. All model development and model maintenance personnel, including data modelers, validation testers, scoring officers, and analysts, can use SAS Model Manager. SAS Model Manager in a business intelligence environment can meet many model development and maintenance challenges.

Here are some of the services SAS Model Manager provides:

- storage of models in a central secure repository.

- creation of custom milestones and tasks to meet business requirements and to match business processes. These milestones and tasks can be used to monitor the development and deployment of models.

- access to data tables that are registered in SAS Management Console or that are located in other SAS libraries for use in SAS Model Manager.

- importing of models into SAS Model Manager. These models can include SAS Enterprise Miner models, SAS/STAT models, PMML models, or models that you develop using SAS code.

- scoring tasks and several reports that can be used to compare, assess, and validate models.

- ability to publish and score models in a specific database using the SAS Scoring Accelerator.

- use of the SAS Integration Technologies Publishing Framework to publish models to a channel.

- use of model performance monitoring and dashboard reports for project champion models and challenger models.

- use of macro programs to run model registration and scoring in a batch environment.

- retraining of models to respond to data or market changes.

For more information, see the *SAS Model Manager: User's Guide*.

SAS OLAP Cube Studio

SAS OLAP Cube Studio enables you to design and create online analytical processing (OLAP) cubes, register cube metadata in a SAS Metadata Repository, and save physical cube data in a specified location. Using the application's Cube Designer Wizard, you can specify the following:

- the data source for a cube

- the cube design and architecture

- measures of the cube for future queries

- initial aggregations for the cube

SAS OLAP Cube Studio also includes functions and wizards that enable you to create additional cube aggregations, to add calculated members to a cube, and to modify existing calculated members.

For more information, see the SAS OLAP Cube Studio Help, which is available from within the product, and the *SAS OLAP Server: User's Guide*. For information about administrative tasks associated with SAS OLAP Cube Studio, see the *SAS Intelligence Platform: Desktop Application Administration Guide*.

SAS Visual Analytics

SAS Visual Analytics is a web-based product that leverages SAS high-performance analytic technologies and complements the capabilities of the SAS Intelligence Platform. It provides an integrated suite of applications that are accessed from a home page. The central entry point enables users to perform a wide variety of tasks such as preparing data sources, exploring data, designing reports, as well as analyzing and interpreting data. Most important, reports can be displayed on a mobile device or on the web.

Business users can explore data on their own to develop new insights, and analysts can speed up the modeling cycle by quickly identifying statistically relevant variables. A key component of SAS Visual Analytics is the SAS LASR Analytic Server, which quickly reads data into memory for fast processing and data visualization.

SAS Visual Analytics provides the following benefits:

- enables users to apply the power of SAS analytics to massive amounts of data

- empowers users to visually explore data, based on any variety of measures, at amazingly fast speeds

- enables users to quickly create powerful statistical models if SAS Visual Statistics is licensed

- enables users to quickly create reports or dashboards using standard tables, graphs, and gauges

- enables users to quickly create customized graphs

- enables users to share insights with anyone, anywhere, via the web or a mobile device

For more information, see http://www.sas.com/software/visual-analytics/overview.html.

SAS Visual Statistics

SAS Visual Statistics extends SAS Visual Analytics and enables you to develop and test models using the in-memory capabilities of SAS software. SAS Visual Analytics Explorer (the explorer) enables you to explore, investigate, and visualize data sources to uncover relevant patterns. SAS Visual Statistics extends these capabilities by creating, testing, and comparing models based on the patterns discovered in the explorer. SAS Visual Statistics can export the score code, before or after performing model comparison, for use with other SAS products and to put the model into production.

SAS Visual Statistics enables you to rapidly create powerful statistical models in a web-based interface. After you have created two or more competing models for your data, SAS Visual Statistics provides a model comparison tool. You can use the tool to create a linear regression model, a logistic regression model, an advanced decision tree, and a clustering tool.

SAS Web Parts for Microsoft SharePoint

SAS Web Parts for Microsoft SharePoint is an integrated set of controls that enable you to provide customized, dynamic content on your website. By using Microsoft Windows SharePoint Services, you can add SAS content directly to your website.

Documentation for SAS Web Parts for Microsoft SharePoint is available from http://support.sas.com/documentation/onlinedoc/webparts.

SAS Web Report Studio

SAS Web Report Studio enables you to create, view, and organize reports. You can use SAS Web Report Studio for the following tasks:

Creating reports
> Beginning with a simple and intuitive view of your data provided by SAS Information Maps (created in SAS Information Map Studio), you can create reports based on either relational or multidimensional data sources. Advanced users can access tables and cubes directly, without the need for information maps.
>
> You can use the Report Wizard to quickly create simple reports, or you can use Edit mode to create sophisticated reports that have multiple data sources, each of which can be filtered. These reports can include various combinations of list tables, crosstabulation tables, graphs, images, and text. Using Edit mode, you can adjust the style to change colors and fonts. You can also insert stored processes, created by business analysts who are proficient in SAS, that contain instructions for calculating analytical results. The stored process results can be rendered as part of a report or as a complete report.
>
> Additional layout and query capabilities are available for advanced users. These users can incorporate custom calculations and complex filter combinations, multiple queries, prompts, and SAS analytical results into a single document. In addition, advanced users can use headers, footers, images, and text to incorporate corporate standards, confidentiality messages, and hyperlinks into reports. Advanced users can also access tables and cubes directly.

Viewing and working with reports
> While viewing reports, you can filter, sort, and rank the data that is shown in list tables, crosstabulation tables, and graphs. With multidimensional data, you can drill down on data in crosstabulation tables and graphs, and drill through to the underlying data.

Organizing reports
> You can create folders and subfolders for organizing your reports. Information consumers can use keywords to find the reports that they need. Reports can be shared with others or kept private. You can schedule reports to run at specified times and distribute them via email, either as PDF attachments or as embedded HTML files. You can also publish reports to one or more publication channels.

Printing and exporting reports

You can preview a report in PDF and print the report, or save and email it later. You have control over many printing options, including page orientation and page range. You can also export data as a spreadsheet and export graphs as images. You can also export data for list tables, crosstabulation tables, and graphs. The output can be viewed in Microsoft Excel or Microsoft Word.

For more information, see the SAS Web Report Studio Help and the *SAS Web Report Studio: User's Guide*, which are available from within the product. For information about administrative tasks associated with SAS Web Report Studio, see the *SAS Intelligence Platform: Web Application Administration Guide*.

SAS Workflow Studio

SAS Workflow Studio enables you to modify and extend the default workflow templates that are provided with certain SAS applications and solutions. SAS Workflow Studio is a modeling tool for the rapid development of workflow templates that reflect an application's business logic. It enables business users to build, organize, and deploy complex workflows, and to adjust business logic efficiently when business demands change. The workflow service layer can be integrated with any number of existing applications.

For more information, see the *SAS Workflow Manager: User's Guide*.

SAS Help Viewer for the Web

The SAS Help Viewer for the Web enables users to view and navigate SAS Online Help in web applications such as the SAS Information Delivery Portal, the SAS BI Dashboard, and SAS Web Report Studio. For information about administration tasks associated with SAS Help Viewer for the Web, see the *SAS Intelligence Platform: Web Application Administration Guide*.

Chapter 7
Security Overview

Overview of SAS Intelligence Platform Security

The security features of the SAS Intelligence Platform offer the following benefits:

- single sign-on from and across disparate systems

- secure access to data and metadata

- role-based access to application features

- confidential transmission and storage of data

- logging and auditing of security events

- .access control reporting

- encryption

The SAS Intelligence Platform's security model cooperates with external systems such as the host environment, the web realm, and third-party databases. To coordinate identity information, SAS keeps a copy of one or more IDs (such as host, Active Directory,

LDAP, or web account IDs) for each user. This requirement does not apply to any users for whom a generic PUBLIC identity is sufficient.

For a comprehensive discussion of security and detailed information about security administration activities, see the *SAS Intelligence Platform: Security Administration Guide* and the *SAS Guide to Metadata-Bound Libraries*, which are available at http://support.sas.com/94administration.

Authorization and Permissions Overview

Metadata-Based Authorization

Authorization is the process of determining which users have which permissions for which resources. The SAS Intelligence Platform includes an authorization mechanism that consists of access controls that you define and store in a metadata repository. These metadata-based controls supplement protections from the host environment and other systems. You can use the metadata authorization layer to manage access to the following resources:

- almost any metadata object (for example, reports, data definitions, information maps, jobs, stored processes, and server definitions)

- OLAP data

- relational data (depending on the method by which the data is accessed)

You can set permissions at several levels of granularity:

- Repository-level controls provide default access controls for objects for which no other access controls are defined.

- Resource-level controls manage access to a specific item such as a report, an information map, a stored process, a table, a column, a cube, or a folder. The controls can be defined individually (as explicit settings) or in patterns (by using access control templates).

- Fine-grained controls affect access to subsets of data within a resource. You can use these controls to specify who can access particular rows within a table or members within a cube dimension.

You can assign permissions to individual users or to user groups. Each SAS user has an identity hierarchy that starts with the user's individual SAS identity and can include multiple levels of nested group memberships.

The effect of a particular permission setting is influenced by any related settings that have higher precedence. For example, if a report inherits a grant from its parent folder but also has an explicit denial, the explicit setting determines the outcome.

The available metadata-based permissions are summarized in the following table.

Table 7.1 *Metadata-Based Permissions*

Permissions	Use
ReadMetadata, WriteMetadata, WriteMemberMetadata, CheckInMetadata	Use to control user interactions with a metadata object.

Permissions	Use
Read, Write, Create, or Delete	Use to control user interactions with the underlying computing resource that is represented by a metadata object; and to control interactions with some metadata objects, such as dashboard objects.
Administer	Use to control administrative interactions (such as starting or stopping) with the SAS server that is represented by a metadata object.

Secured library objects and secured table objects are subject to additional metadata-based permissions.

Metadata-Bound Libraries

To enable you to further control access to physical data, Base SAS includes the ability to define metadata-bound libraries. A metadata-bound library is a physical library that is tied to a corresponding metadata object. Each physical table within a metadata-bound library has information in its header that points to a specific metadata object (a secured table object). The pointer creates a security binding between the physical table and the metadata object.

For SAS9.4M5 a stronger encryption (AES2) is supported for metadata-bound libraries. When you choose to encrypt, you are given a choice of AES or AES2.

The binding ensures that SAS universally enforces metadata-layer permission requirements for the physical table—regardless of how a user requests access from SAS. Users who attempt to reference the data directly (for example, through a LIBNAME statement, when using SAS code, or using tools such as SAS Enterprise Guide) are subject to the same metadata-based authorization as users who request the data through a BI client (such as SAS Web Report Studio).

Locked-down Servers

Another way to control access to physical data is to use locked-down servers. A locked-down server is a SAS server that is allowed to access only specified host resources (directory paths and files). Regardless of host-layer permissions, FILENAME and LIBNAME statements that users submit through a locked-down server are rejected, unless the target resource is included in the server's lockdown paths list.

You can place the following types of servers in a locked-down state: standard and pooled workspace servers, stored process servers, batch servers, grid servers, and SAS/CONNECT servers.

External Authorization Mechanisms

A user's ability to perform a particular action is determined not only by metadata-based and Base SAS mechanisms, but also by external authorization mechanisms such as operating system permissions and database controls. To perform a particular action, the user must have the necessary permissions in all of the applicable authorization layers. For example, regardless of the access controls that have been defined for the user in the metadata repository, the user cannot access a particular file if the operating system permissions do not permit the action.

Roles and Capabilities Overview

The SAS implementation of roles enables administrators to manage the availability of application features such as menu items, plug-ins, and buttons. Applications that have roles include the SAS Add-In for Microsoft Office, SAS Enterprise Guide, SAS Forecast Studio, SAS Management Console, and SAS Web Report Studio. For example, role memberships determine whether a user can see the Server Manager plug-in (in SAS Management Console), compare data (in SAS Enterprise Guide), or directly open an information map (in SAS Web Report Studio). Administrators can assign roles to users and to groups.

An application feature that is under role management is called a capability. Each application that supports roles provides a fixed set of explicit and implicit capabilities. Explicit capabilities can be incrementally added to or removed from any role (other than the unrestricted role, which always provides all explicit capabilities). An implicit capability is permanently bound to a certain role. A contributed capability is an implicit or explicit capability that is assigned through role aggregation. If one role is designated as a contributing role for another role, all of the first role's capabilities become contributed capabilities for the second role.

In general, roles are separate from permissions and do not affect access to metadata or data.

Authentication and Identity Management Overview

Authentication is an identity verification process that attempts to determine whether users (and other entities) are who they say they are. In the simplest case, users already have accounts that are known to the metadata server's host. For example, if the metadata server is on UNIX, then users might have accounts in an LDAP provider that the UNIX host recognizes. If the metadata server is on Windows, then users might have Active Directory accounts.

For accountability, we recommend creating an individual SAS identity for each person who uses the SAS environment. These identities enable administrators to make access distinctions and audit individual actions in the metadata layer. The identities also provide personal folders for each user. The metadata server maintains its own copy of each user ID for the purpose of establishing a SAS identity.

Identity management tasks can be performed manually using SAS Management Console or by using the following batch processes:

- To load user information into the metadata repository, you first extract user and group information from one or more enterprise identity sources. Then you use SAS bulk-load macros to create identity metadata from the extracted information. SAS provides sample applications that extract user and group information and logins from an Active Directory server and from UNIX /etc/passwd and /etc/group files.

- To periodically update user information in the metadata repository, you extract user and group information from your enterprise identity sources and from the SAS metadata. Then you use SAS macros to compare the two sets of data and identify the needed updates. After validating the changes, you use SAS macros to load the updates into the metadata repository.

Note: You cannot use these batch processes to manage passwords. Users can manage their own passwords with the SAS Personal Login Manager.

The metadata identity information is used by the security model's credential management and authorization features. For example, when a user logs on to SAS Data Integration Studio, the metadata server wants to know who the user is so that it can determine which libraries, stored processes, and jobs should be displayed in the desktop client. If a user makes a request in SAS Data Integration Studio to run a job against an Oracle table, the Oracle server wants to know who the user is so that it can determine whether the user has access to the data in the table.

Single Sign-On in the SAS Intelligence Platform

Single Sign-On for SAS Desktop Applications

For desktop applications such as SAS Information Map Studio, SAS Enterprise Guide, SAS Data Integration Studio, SAS OLAP Cube Studio, and SAS Management Console, you can use the following single sign-on features:

- You can enable Integrated Windows authentication so that users do not receive a logon prompt when they launch applications. Integrated Windows authentication is a Microsoft technology that generates and validates Windows identity tokens. All participating clients and servers must authenticate against the same Windows domain (or against domains that trust one another).

- Users can also avoid the initial logon prompt by selecting the option to save their credentials in a connection profile. (This option can be disabled on a site-wide basis.)

Single Sign-On for SAS Web Applications

You can enable web authentication so that users do not receive a logon prompt when they launch SAS web applications such as SAS Web Report Studio and the SAS Information Delivery Portal. In this configuration, SAS web applications use whatever authentication scheme you have set up in your web environment. For example, if your web environment is integrated with a third-party authentication provider, then the SAS web applications participate in that scheme.

Single Sign-On for Data Servers and Processing Servers

Seamless access to SAS Stored Process Servers, SAS OLAP Servers, SAS Content Servers, and SAS Pooled Workspace Servers is provided through SAS token authentication. This mechanism causes participating SAS servers to accept users who are connected to the metadata server. No individual external accounts are required, no user passwords are stored in the metadata, and no reusable credentials are transmitted.

Seamless access to SAS Workspace Servers can be provided through SAS token authentication, Integrated Windows authentication, or credential reuse. With credential reuse, when a user provides credentials in the initial logon dialog box, the credentials are added to the user's in-memory credential cache (user context) and then retrieved when access to the workspace server is required.

You can also use Integrated Windows authentication to provide direct connections to OLAP servers (for example, from a data provider) when there is no active connection to the metadata server.

Overview of Initial Users

During installation, several initial user accounts are created. Some of these accounts are created for all installations, some are optional, and some are created only if certain software components are installed. The required users include the following:

- The SAS Administrator account and the SAS Trusted User Account. These users are generally set up as internal accounts, which exist in metadata but are not known to the host machine. The SAS Administrator account has access to all metadata, regardless of SAS permissions settings. The SAS Trusted User is a privileged service account that can act on behalf of other users when connecting to the metadata server.

- The SAS Spawned Servers account and the SAS Installer account, which must be defined in the operating system of certain server machines. The SAS Spawned Servers account is the initially configured process owner for pooled workspace servers and stored process servers. The SAS Installer Account is used to install and configure SAS software. On UNIX and z/OS systems, this account is also the owner of configuration directories and their contents and is the process owner for items such as the metadata server, the OLAP server, and the object spawner.

Other initial users include the LSF Administrator and LSF User, which are required if Platform Suite for SAS is installed. In addition, the SAS Anonymous Web Service User is an optional account that is used to grant clients access to applicable SAS Web Infrastructure Platform components. Most installations set up this user as an internal account, which exists in metadata but is not known to the host machine.

Encryption Overview

SAS offers encryption features to help you protect information on disk and in transit. When passwords must be stored, they are encrypted or otherwise encoded. Passwords that are transmitted by SAS are also encrypted or encoded. You can choose to encrypt all traffic instead of encrypting only credentials.

You can use an industry standard encryption algorithm for passwords such as AES. For SAS 9.4M5 you can use a new encoding type: SAS005 (AES encryption with 64-bit salt and 10,000 iterations). While the salt value of SAS005 is the same as that of SAS004, SAS005 also has 10,000 iterations, which increases security for stored passwords. You can configure the metadata server to store any new or updated passwords using SAS005.

SAS/SECURE offers maximum protection, including support of the Federal Information Processing Standard (FIPS) 140-2 encryption specification. If you have not installed SAS/SECURE, you can use the SASProprietary algorithm to help protect information.

Note: SAS/SECURE is delivered as part of Base SAS and is included in most licenses.

Security Reporting and Logging Overview

Security reporting creates a snapshot of metadata layer access control settings. SAS provides the %MDSECDS autocall macro to enable you to easily build data sets of permissions information. You can use those data sets as the data source for security reports. You can also identify changes in settings by comparing data sets that are generated at different times.

Security logging records security-related events as part of the system-wide logging facility. The events that are captured include authentication events, client connections, changes to user and group information, and permission changes. For more information about logging, see "Logging and Monitoring for SAS Servers" on page 32 and the *SAS Logging: Configuration and Programming Reference*.

Recommended Reading

Here is the recommended reading list for this title:

- *SAS Intelligence Platform: Installation and Configuration Guide*
- *SAS Intelligence Platform: Migration Guide*
- *SAS Intelligence Platform: System Administration Guide*
- *SAS Intelligence Platform: Security Administration Guide*
- *SAS Intelligence Platform: Application Server Administration Guide*
- *SAS Intelligence Platform: Desktop Application Administration Guide*
- *SAS Intelligence Platform: Data Administration Guide*
- *SAS Intelligence Platform: Middle-Tier Administration Guide*
- *SAS Intelligence Platform: Web Application Administration Guide*
- *SAS Environment Manager User's Guide*
- *Grid Computing in SAS*
- *SAS BI Dashboard: User's Guide*
- *SAS Data Integration Studio: User's Guide*
- *SAS and Hadoop Technology: Overview*
- *SAS High-Performance Forecasting: User's Guide*
- *SAS Information Map Studio: Getting Started with SAS Information Maps*
- *SAS LASR Analytic Server: Reference Guide*
- *SAS Logging: Configuration and Programming Reference*
- *SAS OLAP Server: MDX Guide*
- *SAS OLAP Server: User's Guide*
- *SAS Providers for OLE DB: Cookbook*
- *SAS Scalable Performance Data Engine: Reference*
- *SAS Visual Analytics: User's Guide*
- *SAS Web Report Studio: User's Guide*
- *SAS/ACCESS for Relational Databases: Reference*
- *Scheduling in SAS*

For a complete list of SAS publications, go to sas.com/store/books. If you have questions about which titles you need, please contact a SAS Representative:

SAS Books
SAS Campus Drive
Cary, NC 27513-2414
Phone: 1-800-727-0025
Fax: 1-919-677-4444
Email: sasbook@sas.com
Web address: sas.com/store/books

Glossary

access control template (ACT)

a reusable named authorization pattern that you can apply to multiple resources. An access control template consists of a list of users and groups and indicates, for each user or group, whether permissions are granted or denied.

ACT

See access control template.

Application Response Measurement (ARM)

the name of an application programming interface that was developed by an industry partnership and which is used to monitor the availability and performance of software applications. ARM monitors the application tasks that are important to a particular business.

ARM

See Application Response Measurement.

authentication

See client authentication.

authentication provider

a software component that is used for identifying and authenticating users. For example, an LDAP server or the host operating system can provide authentication.

authorization

the process of determining the permissions that particular users have for particular resources. Authorization either permits or denies a specific action on a specific resource, based on the user's identity and on group memberships.

browser

See web browser.

capability

an application feature that is under role-based management. Typically, a capability corresponds to a menu item or button. For example, a Report Creation capability might correspond to a New Report menu item in a reporting application. Capabilities are assigned to roles.

client authentication (authentication)

the process of verifying the identity of a person or process for security purposes. Authentication is commonly used in providing access to software, and to data that contains sensitive information.

client-side pooling

a configuration in which the client application maintains a collection of reusable workspace server processes.

credential

evidence that is submitted to support a claim of identity (for example, a user ID and password) or privilege (for example, a passphrase or encryption key). Credentials are used to authenticate a user.

cube

See OLAP cube.

data mart

a subset of the data in a data warehouse. A data mart is optimized for a specific set of users who need a particular set of queries and reports.

data set

See SAS data set.

data warehouse (warehouse)

a collection of pre-categorized data that is extracted from one or more sources for the purpose of query, reporting, and analysis. Data warehouses are generally used for storing large amounts of data that originates in other corporate applications or that is extracted from external data sources.

database management system (DBMS)

a software application that enables you to create and manipulate data that is stored in the form of databases. *See also* relational database management system.

database server

a server that provides relational database services to a client. Oracle, DB/2, and Teradata are examples of relational databases.

DBMS

See database management system.

encryption

the conversion of data by the use of algorithms or other means into an unintelligible form in order to secure data (for example, passwords) in transmission and in storage.

foundation services

See SAS Foundation Services.

HTTP (HyperText Transfer protocol)

a protocol for transferring data to the Internet. HTTP provides a way for servers and web clients to communicate. It is based on the TCP/IP protocol.

HyperText Transfer protocol

See HTTP.

identity

See metadata identity.

information map

a collection of data items and filters that provides a user-friendly view of a data source. When you use an information map to query data for business needs, you do not have to understand the structure of the underlying data source or know how to program in a query language.

Integrated Object Model (IOM)

the set of distributed object interfaces that make SAS software features available to client applications when SAS is executed as an object server.

Integrated Object Model server (IOM server)

a SAS object server that is launched in order to fulfill client requests for IOM services.

Integrated Windows authentication (IWA)

a Microsoft technology that facilitates use of authentication protocols such as Kerberos. In the SAS implementation, all participating components must be in the same Windows domain or in domains that trust each other.

internal account

a SAS account that you can create as part of a user definition. Internal accounts are intended for metadata administrators and some service identities; these accounts are not intended for regular users.

IOM

See Integrated Object Model.

IOM server

See Integrated Object Model server.

IWA

See Integrated Windows authentication.

LDAP (Lightweight Directory Access Protocol)

a protocol that is used for accessing directories or folders. LDAP is based on the X. 500 standard, but it is simpler and, unlike X.500, it supports TCP/IP.

Lightweight Directory Access Protocol

See LDAP.

load balancing

for IOM bridge connections, a program that runs in the object spawner and that uses an algorithm to distribute work across object server processes on the same or separate machines in a cluster.

logical server

the second-level object in the metadata for SAS servers. A logical server specifies one or more of a particular type of server component, such as one or more SAS Workspace Servers.

MDDB

See multidimensional database.

metadata

descriptive data about data that is stored and managed in a database, in order to facilitate access to captured and archived data for further use.

metadata identity (identity)

a metadata object that represents an individual user or a group of users in a SAS metadata environment. Each individual and group that accesses secured resources on a SAS Metadata Server should have a unique metadata identity within that server.

metadata LIBNAME engine

the SAS engine that processes and augments data that is identified by metadata. The metadata engine retrieves information about a target SAS library from metadata objects in a specified metadata repository.

metadata object

a set of attributes that describe a table, a server, a user, or another resource on a network. The specific attributes that a metadata object includes vary depending on which metadata model is being used.

metadata server

a server that provides metadata management services to one or more client applications.

multidimensional database (MDDB)

a specialized data storage structure in which data is presummarized and cross-tabulated and then stored as individual cells in a matrix format, rather than in the row-and-column format of relational database tables. The source data can come either from a data warehouse or from other data sources. MDDBs can give users quick, unlimited views of multiple relationships in large quantities of summarized data.

object spawner (spawner)

a program that instantiates object servers that are using an IOM bridge connection. The object spawner listens for incoming client requests for IOM services.

OLAP (online analytical processing)

a software technology that enables users to dynamically analyze data that is stored in multidimensional database tables (cubes).

OLAP cube (cube)

a logical set of data that is organized and structured in a hierarchical, multidimensional arrangement to enable quick analysis of data. A cube includes measures, and it can have numerous dimensions and levels of data.

OLAP schema

a container for OLAP cubes. A cube is assigned to an OLAP schema when it is created, and an OLAP schema is assigned to a SAS OLAP Server when the server is defined in the metadata. A SAS OLAP Server can access only the cubes that are in its assigned OLAP schema.

online analytical processing

See OLAP.

parallel execution

See parallel processing.

parallel I/O

a method of input and output that takes advantage of multiple CPUs and multiple controllers, with multiple disks per controller to read or write data in independent threads.

parallel processing (parallel execution)

a method of processing that divides a large job into multiple smaller jobs that can be executed simultaneously on multiple CPUs.

permission

the type of access that a user or group has to a resource. The permission defines what the user or group can do with the resource. Examples of permissions are ReadMetadata and WriteMetadata.

plug-in

a file that modifies, enhances, or extends the capabilities of an application program. The application program must be designed to accept plug-ins, and the plug-ins must meet design criteria specified by the developers of the application program.

RDBMS

See relational database management system.

relational database management system (RDBMS)

a database management system that organizes and accesses data according to relationships between data items. The main characteristic of a relational database management system is the two-dimensional table. Examples of relational database management systems are DB2, Oracle, Sybase, and Microsoft SQL Server.

role

See user role.

SAS Application Server

a logical entity that represents the SAS server tier, which in turn comprises servers that execute code for particular tasks and metadata objects.

SAS Content Server

a server that stores digital content (such as documents, reports, and images) that is created and used by SAS client applications. To interact with the server, clients use WebDAV-based protocols for access, versioning, collaboration, security, and searching.

SAS data set (data set)

a file whose contents are in one of the native SAS file formats. There are two types of SAS data sets: SAS data files and SAS data views.

SAS Foundation Services (foundation services)

a set of core infrastructure services that programmers can use in developing distributed applications that are integrated with the SAS platform. These services provide basic underlying functions that are common to many applications. These functions include making client connections to SAS application servers, dynamic service discovery, user authentication, profile management, session context management, metadata and content repository access, information publishing, and stored process execution. *See also* service.

SAS Management Console

a Java application that provides a single user interface for performing SAS administrative tasks.

SAS Metadata Repository

a container for metadata that is managed by the SAS Metadata Server. *See also* SAS Metadata Server.

SAS Metadata Server

a multi-user server that enables users to read metadata from or write metadata to one or more SAS Metadata Repositories.

SAS OLAP Server

a SAS server that provides access to multidimensional data. The data is queried using the multidimensional expressions (MDX) language.

SAS Open Metadata Architecture

a general-purpose metadata management facility that provides metadata services to SAS applications. The SAS Open Metadata Architecture enables applications to exchange metadata, which makes it easier for these applications to work together.

SAS Scalable Performance Data Server (SPD Server)

a server that restructures data in order to enable multiple threads, running in parallel, to read and write massive amounts of data efficiently.

SAS Stored Process (stored process)

a SAS program that is stored on a server and defined in metadata, and which can be executed by client applications.

SAS Stored Process Server

a SAS IOM server that is launched in order to fulfill client requests for SAS Stored Processes.

SAS token authentication

a process in which the metadata server generates and verifies SAS identity tokens to provide single sign-on to other SAS servers. Each token is a single-use, proprietary software representation of an identity.

SAS Web Infrastructure Platform

a collection of middle-tier services and applications that provide infrastructure and integration features that are shared by SAS web applications and other HTTP clients.

SAS Workspace Server

a SAS server that provides access to SAS Foundation features such as the SAS programming language and SAS libraries.

SAS/SHARE server

the result of an execution of the SERVER procedure, which is part of SAS/SHARE software. A server runs in a separate SAS session that services users' SAS sessions by controlling and executing input and output requests to one or more SAS libraries.

SASProprietary algorithm

a fixed encoding algorithm that is included with Base SAS software. The SASProprietary algorithm requires no additional SAS product licenses. It provides a medium level of security.

server-side pooling

a configuration in which a SAS object spawner maintains a collection of reusable workspace server processes that are available for clients. The usage of servers in this pool is governed by the authorization rules that are set on the servers in the SAS metadata.

service

one or more application components that an authorized user or application can call at any time to provide results that conform to a published specification. For example, network services transmit data or provide conversion of data in a network, database services provide for the storage and retrieval of data in a database, and web services interact with each other on the World Wide Web. *See also* SAS Foundation Services.

single sign-on (SSO)

an authentication model that enables users to access a variety of computing resources without being repeatedly prompted for their user IDs and passwords. For example, single sign-on can enable a user to access SAS servers that run on different platforms without interactively providing the user's ID and password for each platform. Single sign-on can also enable someone who is using one application to launch other applications based on the authentication that was performed when the user initially logged on.

SMP (symmetric multiprocessing)

a type of hardware and software architecture that can improve the speed of I/O and processing. An SMP machine has multiple CPUs and a thread-enabled operating system. An SMP machine is usually configured with multiple controllers and with multiple disk drives per controller.

spawner

See object spawner.

SPD Server

See SAS Scalable Performance Data Server.

SSO

See single sign-on.

stored process

See SAS Stored Process.

symmetric multiprocessing

See SMP.

theme

a collection of specifications (for example, colors, fonts, and font styles) and graphics that control the appearance of an application.

thread

the smallest unit of processing that can be scheduled by an operating system.

transformation

in data integration, an operation that extracts data, transforms data, or loads data into data stores.

user role (role)

a set of capabilities within an application that are targeted to a particular group of users.

warehouse

See data warehouse.

web application

an application that is accessed via a web browser over a network such as the Internet or an intranet. SAS web applications are Java Enterprise Edition (JEE) applications that are delivered via web application archive (WAR) files. The applications can depend on Java and non-Java web technologies.

web authentication

a configuration in which users of web applications and web services are verified at the web perimeter, and the metadata server trusts that verification.

web browser (browser)

a software application that is used to view web content, and also to download or upload information. The browser submits URL (Uniform Resource Locator) requests to a web server and then translates the HTML code into a visual display.

Index

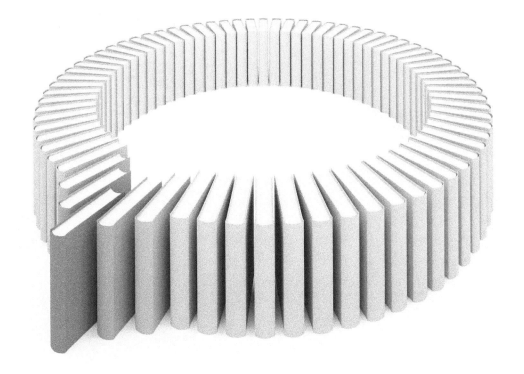

Gain Greater Insight into Your SAS® Software with SAS Books.

Discover all that you need on your journey to knowledge and empowerment.

 support.sas.com/bookstore
for additional books and resources.

9 781629 600895